T0261291

NATURAL

COMPLEXITY

PRIMERS IN COMPLEX SYSTEMS

VOLUMES PUBLISHED IN THE SERIES

Natural Complexity: A Modeling Handbook,
by Paul Charbonneau (2017)

Spin Glasses and Complexity,
by Daniel L. Stein and Charles M. Newman (2013)

Diversity and Complexity, by Scott E. Page (2011)

Phase Transitions, by Ricard V. Solé (2011)

Ant Encounters: Interaction Networks and Colony Behavior,
by Deborah M. Gordon (2010)

NATURAL COMPLEXITY

A Modeling Handbook

Paul Charbonneau

PRINCETON UNIVERSITY PRESS

Princeton & Oxford

Published by Princeton University Press, 41 William Street,
Princeton, New Jersey 08540
In the United Kingdom: Princeton University Press, 6 Oxford Street,
Woodstock, Oxfordshire OX20 1TR

press.princeton.edu

Cover image courtesy of Don Komarechka Photography

ISBN: 978-0-691-17684-0
ISBN (pbk.): 978-0-691-17035-0

Library of Congress Control Number: 2016953537

British Library Cataloging-in-Publication Data is available

This book has been composed in Adobe Garamond and Helvetica Neue
Printed on acid-free paper. ∞

Typeset by Nova Techset Pvt Ltd, Bangalore, India
Printed in the United States of America

1 3 5 7 9 10 8 6 4 2

To Jrène, Roland, and Nico,
key agents in a little complex system I could not do without

CONTENTS

PREFACE

A preface seldom makes for exciting reading, as the bulk of its square footage is typically used to thank people the vast majority of readers have never heard of and never will meet. Nonetheless, anyone going through the process of publishing a book accumulates debts that must be acknowledged. So let's first do this, swiftly but in true heartfelt and thankful spirit, and then move on with the real stuff.

The material in many of the book's chapters grew from class notes and end-of-term projects I developed for the first-year undegraduate course "Introduction to computational physics," offered at the Physics Department of the Université de Montréal, and which I had the opportunity to teach for many years. I am grateful for constructive feedback by generations of students and colleagues, far too numerous to list in full. A few went above and beyond the call of duty, in some cases years after having taken the class, as these notes evolved into a book and the associated codes went from C to Python. This is why I do feel an irresistible urge to thank by name: Vincent Aymong, Vincent Dumoulin, Mirjam Fines-Neuschild, David Lafrenière, Myriam Latulippe, Guillaume Lenoir-Craig, Dário Passos, Antoine Strugarek, and Félix Thouin.

Expanding these class notes and term projects into the first draft of the present book was made possible by a one-year paid sabbatical leave for which I should at the very least express gratitude to my employer. Turning that first draft into the actual book you now have in hand involved a lot of hard work by many people at Princeton University Press, including in the front row Ingrid Gnerlich, Eric Henney, and Kathleen Cioffi. Those in the other rows I never interacted with directly, but I know you are there somewhere, so thank you too. I am particularly indebted to Alison Durham for her outstanding copyediting work, Don Komarechka and Ken Libbrecht for sharing their snowflake photographs; and last but certainly not least, my family for putting up with me during the more intense chapters of this saga.

Finally, I wish to acknowledge many valuable suggestions and constructive criticism by the two anonymous readers appointed by Princeton University Press to review the first draft of this book. Needless to say, any remaining error, inaccuracy, or omission is not intentional, and neither one of these two individuals nor any other named earlier holds any responsibility for their (probable) occurrence; but I do, so if you find any, please let me know so I can do something about it next time around, if there is one.

Paul Charbonneau
Orford, November 2016

NATURAL

COMPLEXITY

1

INTRODUCTION: WHAT IS COMPLEXITY?

There is all the difference in the world between *knowing about* and *knowing how to do*.
— J. EVANS, *The History and Practice of Ancient Astronomy*, 1997)

1.1 Complexity Is Not Simple

If turbulence is the graveyard of theories, then complexity is surely the tombstone of definitions. Many books on complexity have been written, and the bravest of their authors have attempted to define complexity, with limited success. Being nowhere as courageous, I have simply decided not to try. Although complexity is the central topic of this book, I hereby pledge to steer clear of any attempt to formally define it.

This difficulty in formally defining complexity is actually surprising because we each have our own intuitive definition of what is "complex" and what is not, and we can usually decide pretty quickly if it is one or the other. To most people a Bartok string quartet "sounds" complex, and a drawing by Escher "looks" complex. Such intuitive definitions can even take an egocentric flavor, i.e., an Escher drawing is complex because "I could not draw it" or a Mozart piano piece is complex because "I could not play it."

The many guises of the complex systems to be encountered further in this book often involve many (relatively) simple individual elements interacting locally with one another. This characterization—it should definitely not be considered a definition—does capture a surprisingly wide range of events, structures, or phenomena occurring in the natural world, that most of us would intuitively label as complex. It even applies to many artificial constructs and products of the human mind. While novels by Thomas Pynchon are typically replete with oddball characters, events therein are for the most part constrained by the laws of physics and usually follow a relatively straightforward timeline. What makes Pynchon's novels complex is that they involve many, many such characters interacting with one another. The complexity arises not from the characters themselves, however singularly they may behave, but rather from their mutual interactions over time. Likewise, many of Escher's celebrated drawings[1] are based on the tiling of relatively simple pictorial elements, which undergo slow, gradual change across the drawing. The complexity lies in the higher-level patterns that arise globally from the mutual relationship of neighboring pictorial units, which are themselves (relatively) simple.

Nice and fine perhaps, but turning this into a formal definition of complexity remains an open challenge. One can turn the problem on its head by coming up instead with a definition of what is *not* complex, i.e., a formal definition of "simple." Again purely intuitive and/or egocentric definitions are possible, such as "simple = my five-year-old could do this." Like complexity, simplicity is, to a good part, in the eye of the beholder. I am a physicist by training and an astrophysicist and teacher by trade; I am well aware that my own personal definition of what is "simple" does not intersect fully with that of most people I know. Yet such

---

[1] See www.mcescher.com/gallery/transformation-prints for reproductions of artwork by Maurits Cornelis Escher.

divergences of opinions are often grounded in the language used to describe and characterize a phenomenon.

Consider, for example, the game of pool.[2] Even without any formal knowledge of energy and momentum conservation, a beginner develops fairly rapidly a good intuitive feel for *how* the cue ball *should* be hit to propel a targeted, numbered ball into a nearby pocket; reliably executing the operation is what requires skill and practice. Now, armed with Newton's laws of motion, and knowing the positions of the pocket and the two participating balls, the required impact point of the cue ball can be *calculated* to arbitrarily high accuracy; the practical problem posed by the production of the proper trajectory of the cue ball, of course, remains. Whichever way one looks at it, the collision of two (perfectly spherical) pool balls is definitely simple, provided it takes place on a perfectly flat table.

If physical laws allow, in principle, the computation of the exact trajectories of two colliding pool balls, the same laws applied repeatedly should also allow generalization to many balls colliding in turn with one another. Experience shows that the situation rapidly degrades as the number of balls increases. I have not played pool much, but that is still enough to state confidently that upon starting the game, no single pool break is ever exactly alike another, despite the fact that the initial configuration of the 15 numbered balls (the "rack") is always the same and geometrically regular—closely packed in a triangular shape. The unfolding of the break depends not just on the speed, trajectory angle, and impact position of the cue ball, but also on the exact distances between the balls in the rack and on whether one ball actually touches another, i.e., on the *exact* position of each ball. For all practical purposes, the break is unpredictable because it exhibits

---

[2]The reader unfamiliar with this game will find, on the following Wikipedia page, just enough information to make sense of the foregoing discussion: en.wikipedia.org/wiki/Eight-ball.

extreme sensitivity to the initial conditions, even though the interaction between any pair of colliding balls is simple and fully deterministic.

Is complexity then just a matter of sheer number? If the definition of complexity is hiding somewhere in the interactions between many basic elements, then at least from a modeling point of view we may perhaps be in business. If the underlying physical laws are known, computers nowadays allow us to *simulate* the evolution of systems made up of many, many components, to a degree of accuracy presumably limited only by the number of significant digits with which numbers are encoded in the computer's memory. This "brute-force" approach, as straightforward as it may appear in principle, is plagued by many problems, some purely practical but others more fundamental. Looking into these will prove useful to start better pinning down what complexity is not.

1.2 Randomness Is Not Complexity

If we are to seriously consider the brute-force approach to the modeling of complex systems, we first need to get a better feel for what is meant by "large number." One simple (!) example should suffice to quantify this important point.

Consider a medium-size classroom, say a 3 m high room with a 10×10 m floor. With air density at $\rho = 1.225 \, \text{kg} \, \text{m}^{-3}$, this 300 m^3 volume contains 367 kg of N_2 and O_2 molecules, adding up to some 10^{28} individual molecules. Written out long that number is

$$10,000,000,000,000,000,000,000,000,000.$$

It doesn't look so bad, but this is actually a *very* large number, even by astronomical standards; just consider that the total number of stars in all galaxies within the visible universe is estimated to be in the range 10^{22}–10^{24}. Another way to appreciate the sheer numerical magnitude of 10^{28} is to reflect upon the fact that 10^{28} close-packed sand grains of diameter 0.25 mm—"medium-grade sand" according to the ISO 14688 standard, but quality beach stuff nonetheless—would cover the whole surface of the Earth, oceans included, with a sandy layer 1 km thick. That is how many molecules we need to track—positions and velocities—to "simulate" air in our classroom.

At this writing, the supercomputers with the largest memory can hold up to $\sim 10^3\,\text{TB} = 10^{15}\,\text{bytes}$ in RAM. Assuming 64-bit encoding of position and velocity components, each molecule requires 48 bytes, so that at most 2×10^{13} molecules can be followed "in-RAM."[3] This is equivalent to a cubic volume element of air smaller than a grain of very fine sand. We are a long way from simulating air in our classroom, and let's not even think about weather forecasting! This is a frustrating situation: we know the physical laws governing the motion and interaction of air molecules, but don't have the computing power needed to apply them to our problem.

Now, back to reality. No one in their right mind would seriously advocate such a brute-force approach to atmospheric modeling, even if it were technically possible, and not only because brute force is seldom the optimal modeling strategy. Simply put, complete detailed knowledge of the state of motion of every single air molecule in our classroom is just not useful in practice. When I walk into a classroom, I am typically interested in global measures such as temperature, humidity level, and perhaps the concentration gradient of Magnum 45 aftershave, so as to pinpoint the location of the source and expel the offending emitter.

It is indeed possible to describe, understand, and predict the behavior of gas mixtures, such as air, through the statistical

---

[3] Molecules also have so-called "internal" degrees of freedom, associated with vibrational and rotational excitation, but for the sake of the present argument these complications can be safely ignored.

definition of global measures based on the physical properties of individual molecules and of the various forces governing their interactions. This statistical approach stands as one of the great successes of nineteenth-century physics. Once again, a simple example can illustrate this point.

The two panels atop figure 1.1 display two different realizations of the spatially random distribution of $N = 300$ particles within the unit square. Even though the horizontal and vertical coordinates of each particle are randomly drawn from a uniform distribution in the unit interval, the resulting spatial distributions are not spatially homogeneous, showing instead clumps and holes, which is expected, considering the relatively small number of particles involved. Viewing these two distributions from a distance, the general look is the same, but comparing closely the two distributions differ completely in detail—not one single red particle on the left is at *exactly* the same position as any single green particle on the right.

Consider now the following procedure: from the center of each unit square, draw a series of concentric circles with increasing radii r; the *particle number density* (ρ, in units of particles per unit area) can be computed by counting the number of particles within each such circle, and dividing by its surface area πr^2. Mathematically, this would be written as

$$\rho(r) = \frac{1}{\pi r^2} \sum_{n=1}^{N} \begin{cases} 1 & \text{if } x_n^2 + y_n^2 \leq r^2, \\ 0 & \text{otherwise.} \end{cases} \tag{1.1}$$

Clearly, as the radius r is made larger, more and more particles are contained within the corresponding circles, making the sum in equation (1.1) larger, but the area πr^2 also increases, so it is not entirely clear a priori how the density will vary as the radius r increases. The bottom panel of figure 1.1 shows the results of this exercise, applied now to two realizations of not 300 but $N = 10^7$ particles, again randomly distributed in the unit

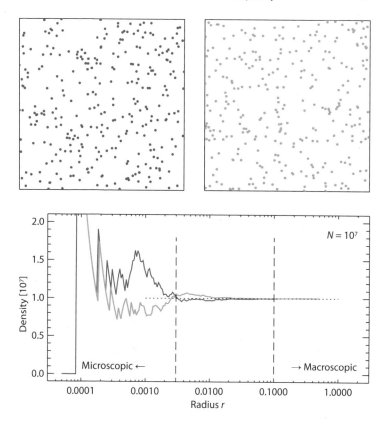

Figure 1.1. Going from the microscopic to the macroscopic scale. The top panels show two distinct random distributions of $N = 300$ particles in the unit square. The bottom panel shows the result of using equation (1.1) to calculate the particle density, based on a series of circles of increasing radii, concentric and centered on the middle of the unit square, now for two distinct random distributions of $N = 10^7$ particles. Note the logarithmic horizontal axis. The resulting density curves differ completely for radii smaller than a few times the mean interparticle distance $\delta = 0.0003$, but converge to the expected value of 10^7 particles per unit area for radii much larger than this distance (see text).

square. The statistically uniform packing of $N = 10^7$ particles in the unit square implies a typical interparticle distance of order $\delta \simeq 1/\sqrt{N} \sim 0.0003$ here. For radii r of this order or smaller, in equation (1.1) the computed density value is critically dependent on the exact position of individual particles, and for $r < \delta$ is it quite possible that no particle is contained within the circle, leading to $\rho = 0$. This is what is happening for the red curve in figure 1.1 up to $r \simeq 0.0001$, while in the case of the distribution associated with the green curve it just so happens that a clump of particles is located at the center of the unit square, leading to abnormally large values for the density even for radii smaller than δ. Nonetheless, as r becomes much larger than δ, both curves converge to the expected value $\rho = 10^7$ particles per unit area.

Figure 1.1 illustrates a feature that will be encountered repeatedly in subsequent chapters of this book, namely, *scale separation*. At the *microscopic* scale (looking at the top panels of figure 1.1 up close) individual particles can be distinguished, and the description of the system requires the specification of their positions, and eventually their velocities and internal states, if any. In contrast, at the *macroscopic scale* (looking at the top panels of figure 1.1 from far back), global properties can be defined that are independent of details at the microscopic scale. Of course, if two systems are strictly identical at the microscopic level, their global properties will also be the same. What is more interesting is when two systems differ at the microscopic level, such as in the two top panels of figure 1.1, but have the same statistical properties (here, x and y coordinates are uniformly distributed in the unit interval); then their physical properties at the macroscopic scale, such as density, will also be the same.

It is worth reflecting a bit more upon this whole argument in order to fully appreciate under which conditions global properties such as density can be meaningfully defined. Considering the statistical nature of the system, one may be tempted to conclude that what matters most is that N be large; but what do we mean by "large"? Large with respect to what? The crux is really that a good separation of scale should exist between the microscopic and macroscopic. The interparticle distance δ (setting the microscopic scale) must be much smaller than the macroscopic scale L at which global properties are defined; in other words, N should be large enough so that $\delta \ll L$. The two vertical dashed lines in figure 1.1 have been drawn to indicate the scale boundaries of the microscopic and macroscopic regimes; the exact values of r chosen are somewhat arbitrary, but a good separation of scale implies that these two boundaries should be as far as possible from one another. In the case of the air in our hypothetical classroom, $\delta \simeq 3 \times 10^{-9}$ m, so that with a macroscopic length scale ~ 1 m, scale separation is very well satisfied.

What happens in the intermediate scale regime, i.e., between the two dashed lines in figure 1.1, is an extremely interesting question. Typically, meaningful global properties cannot be defined, and N is too large to be computationally tractable as a direct simulation. In closed thermodynamic systems (such as the air in our classroom), also lurking somewhere in this twilight zone of sorts is the directionality of time: (elastic) collisions between any two molecules are entirely time reversible, but macroscopic behavior, such as the spread of olfactorily unpleasant aftershave molecules from their source, is not, even though it ultimately arises from time-reversible collisions. Fascinating as this may be, it is a different story, so we should return to complexity since this is complex enough already.

If large N and scale separation are necessary conditions for the meaningful definition of macroscopic variables, they are not sufficient conditions. In generating the two top panels of figure 1.1, particles are added one by one by drawing random numbers from the unit interval to set their horizontal (x) and vertical coordinates (y). The generation of the (x, y) coordinates for a given particle is entirely independent of the positions of the particles already placed in the unit square; particle positions are entirely *uncorrelated*. In subsequent chapters, we will repeatedly encounter situations where the "addition" of a particle to a system is entirely set by the locations of the particles already in the system. Particle positions are then strongly correlated, and through these correlations complexity can persist at all scales up to the macroscopic.

To sum up the argument, while systems made up of many interacting elements may appear quite complex at their microscopic scale, there are circumstances under which their behavior at the macroscopic scale can be subsumed into a few global quantities for which simple evolutionary rules can be constructed or inferred experimentally. The take-home message here then is the following: although complex natural systems often involve a large number of (relatively) simple individual elements interacting locally with one another, not all systems made up of many interacting elements exhibit complexity in the sense to be developed throughout this book. The 10^{28} air molecules in our model classroom, despite their astronomically large number and ever-occurring collisions with one another, collectively add up to a simple system.

1.3 Chaos Is Not Complexity

Complex behavior can actually be generated in systems of very few interacting elements. Chaotic dynamics is arguably the best known and most fascinating generator of such behavior, and there is no doubt that patterns and structures produced by systems exhibiting chaotic dynamics are "complex," at least in the intuitive sense alluded to earlier.

Practically speaking, generators of chaotic dynamics can be quite simple indeed. The *logistic map*, a very simple model of population growth under limited carrying capacity of the environment, provides an excellent case in point. Consider a biological species with a yearly reproduction cycle, and let x_n measure the population size at year n. Under the logistic model of population growth, the population size at year $n + 1$ is given by

$$x_{n+1} = A\, x_n (1 - x_n), \quad n = 0, 1, 2, \ldots, \quad (1.2)$$

where A is a positive constant, and x_0 is the initial population. Depending on the chosen numerical value of A, the iterative sequence $x_0, x_1, x_2, \ldots$ can converge to zero, or to a fixed value, or oscillate periodically, multiperiodically, or aperiodically as a function of the iteration number n. These behaviors are best visualized by constructing a *bifurcation diagram*, as in the bottom-left panel of figure 1.2. The idea is to plot successive values of x_n produced with a given value of A, excluding, if needed, the transient phase during which the initial value x_0 converges to its final value or set of values, and repeating this process for progressively larger values of A. Here, for values of $1.0 < A < 3.0$, the iterative sequence converges to a fixed nonzero numerical value, which gradually increases with increasing A; this leads to a slanted line in the bifurcation diagram, as successive values of x_n for a given A are all plotted atop one another. Once A exceeds 3.0, the iterates alternate between two values, leading to a split into two branches in the bifurcation diagram. Further increases in A lead to successive splittings of the various branches, until the chaotic regime is reached, at which point the iterate x_n varies aperiodically. This is a classical example of transition to chaos through a period-doubling cascade.

The bifurcation diagram for the logistic map is certainly complex in the vernacular sense of the word; most people would certainly have a hard time drawing it with pencil and paper. There is in fact much more to it than that. The series of nested close-ups in figure 1.2 zooms in on the end point of the period-doubling cascade, on a branch of the primary transition to chaos. No matter the zooming level, the successive bifurcations have the same shape and topology. This *self-similarity* is the hallmark of scale invariance, and marks the bifurcation diagram as a fractal

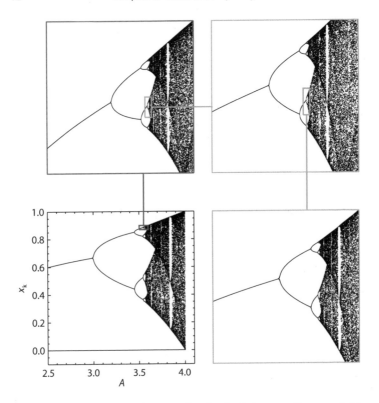

Figure 1.2. Bifurcation diagram for the logistic map (bottom left), as given by equation (1.2). The first bifurcation from the trivial solution $x_n = 0$ occurs at $A = 1.0$, off to the left on the horizontal scale. The other three frames show successive nested close-ups (red→blue→green) on the period-doubling cascade to chaos.

structure. We will have a lot more to say on scale invariance and fractals in subsequent chapters, as these also arise in the many complex systems to be encountered throughout this book.

Chaotic systems such as the logistic map also exhibit *structural sensitivity*, in the sense that they can exhibit qualitative changes of behavior when control parameters—here the numerical constant A—undergo small variations. For example, in the case of the logistic map, increasing A beyond the value 3.0 causes the iterate x_n to alternate between a low and a high value, whereas before, it converged to a single numerical value. In the chaotic regime the map is also characterized by sensitivity to initial conditions, in that the numerical difference between the x_n's of two sequences, differing by an infinitesimally small amount at $n = 0$, is amplified exponentially in subsequent iterations.

Many complex systems to be encountered in the following chapters exhibit similar sensitivities, but for entirely different reasons, usually associated with the existence of long-range correlations established within the system in the course of its prior evolution, through simple and local interactions between their many constitutive elements. In contrast, the cleanest examples of chaotic systems involve a few elements (or degrees of freedom), subject to strong nonlinear coupling. Although such chaotic systems generate patterns and behavior that are complex in the intuitive sense of the word, in and of themselves they are not complex in the sense to be developed in this book.

1.4 Open Dissipative Systems

One common feature of systems generating complexity is that they are *open* and *dissipative*. Pool can serve us well once again in providing a simple example of these notions. After a pool break, the moving balls eventually slow down to rest (with hopefully at least one falling into a pocket in the process). This occurs because of kinetic energy loss due to friction on the table's carpet, and not-quite-elastic collisions with the table's bumpers. The system jointly defined by the moving balls is *closed* because it is subjected to no energy input after the initial break, and is *dissipative* because that energy is slowly lost to friction (and ultimately, heat) until the system reaches its lowest energy equilibrium state: all balls at rest.

Imagine now that the pool table in located inside a ship sailing a rough sea, so that the table is ever slowly and more or less randomly tilted back and forth. Following the break, the balls may slow down to some extent, but will not come to rest since they intermittently pick up energy from the moving table. They will also sometimes temporarily lose kinetic energy of course, for example when finding themselves moving "uphill" due to an unfavorable tilt of the table. But the point is that the balls will not stop moving (well, until they all end up in pockets) no matter how long we wait. A player somehow unaware of the ship's rock-and-roll would undoubtedly wonder at the curiously curved trajectories and spontaneous acceleration and deceleration of the moving pool balls, and perhaps conclude that their seventh piña colada was one too many.

In this seafaring pool situation, the equilibrium state is one where, on average, the table's motion injects energy into the system at the same rate as it gets dissipated into heat by friction. The system is still dissipative but is now also *open*, in that it benefits from an input of energy from an external source. At equilibrium, there is as much energy entering the system as is being dissipated, but the equilibrium state is now more interesting: the balls are perpetually moving and colliding, a consequence of energy moving *through* the system.

A most striking property of open dissipative systems is their ability to generate large-scale structures or patterns persisting far longer than the dynamical timescales governing the interactions of microscopic constituents. A waterfall provides a particularly simple example; it persists with its global shape unchanged for times much, much longer that the time taken by an individual water molecule passing through it. As a physical object, the waterfall is obviously "made up" of water molecules, but as a spatiotemporal structure the identity of its individual water molecules is entirely irrelevant. Yet, block off the water supply upstream, and the waterfall disappears on the (short) timescale it takes a water molecule to traverse it. The waterfall persists as a structure only because water flows through it, i.e., the waterfall is an open system.

This line of argument carries over to systems far more intricate than a "simple" waterfall. Consider, for example, Earth's climate; now that is certainly a complex system in any sense of the word. Climate collects a very wide range of phenomena developing on an equally wide range of spatial and temporal scales: the seasonal cycle, large-scale atmospheric wind patterns such as the jet stream, oceanic currents, recurrent global patterns such as El Niño, tropical storms, and down in scale to thunderstorms and tornadoes, to name but a few. Solar radiative energy entering the atmosphere from above is the energy source ultimately powering all these phenomena. Yet, globally the Earth remains in thermal equilibrium, with as much energy absorbed on the dayside as is radiated back into space over its complete surface in the course of a day. Earth is an open system, with solar energy flowing in and out. If the Sun were to suddenly stop shining, the pole–equator temperature gradient would vanish and all atmospheric and oceanic fluid motions would inexorably grind to a halt, much like the pool balls eventually do after a break on a fixed table. Everything we call climate is just a temporary channeling of a small part of the "input" solar radiative energy absorbed by Earth, all ultimately liberated as heat via viscous dissipation and radiated back into space. The climate maintains its complexity, and generates persistent large-scale weather patterns—the equivalent of our waterfall—by tapping into the energy flowing *through* Earth's atmosphere, surface, and oceans. Earth is an open dissipative system on a very grand scale.

Most complex systems investigated in this book, although quite simple in comparison to Earth's climate, are open dissipative systems in the same sense. They benefit from an outside source of energy, and include one or more mechanisms allowing energy to be evacuated at their boundary or to be dissipated internally (or both).

1.5 Natural Complexity

Although I have wriggled away from formally defining complexity, considering the title of this book I do owe it to the reader to at least clarify what I mean by *natural complexity*, and how this relates to complexity in general.

Exquisitely complex phenomena can be produced in the laboratory under well-controlled experimental conditions. In the field of physics alone, phase transitions and fluid instabilities offer a number of truly spectacular examples. In contrast, the systems investigated throughout this book are idealizations of naturally occurring phenomena characterized by the autonomous generation of structures and patterns at macroscopic scales that are not directed or controlled at the macroscopic level or by some agent external to the system, but arise instead "naturally" from dynamical interactions at the microscopic level. This is one mouthful of a characterization, but it does apply to natural phenomena as diverse as avalanches, earthquakes, solar flares, epidemics, and ant colonies, to name but a few.

Each chapter in this book presents a simple (!) computational model of such natural complex phenomena. That natural complexity can be studied using simple computer-based models may read like a compounded contradiction in terms, but in fact it is not, and this relates to another keyword in this book's title: *modeling*. In the sciences we make models—whether in the form of mathematical equations, computer simulations, or laboratory experiments—in order to isolate whatever phenomenon is of interest from secondary "details," so as to facilitate our understanding of the said phenomenon. A good model is seldom one which includes as much detail as possible for the system under study, but is instead one just detailed enough to answer our specific questions regarding the phenomenon of interest. Modeling is thus a bit of an art, and it is entirely legitimate to construct distinct models of the same given phenomenon, each aimed at understanding a distinct aspect.

To many a practicing geologist or epidemiologist, the claim that the very simple computational models developed in the following chapters have anything to do with real earthquakes or real epidemics may well be deemed professionally offensive, or at best dismissed as an infantile nerdy joke. Such reactions are quite natural, considering that, still today, in most hard sciences explanatory frameworks tend to be strongly reductionist, in the sense that explanations of global behaviors are sought and formulated preferentially in terms of laws operating at the microscopic level. My own field of enquiry, physics, has in fact pretty much set the standard for this approach. In contrast, in the many complex systems modeled in this book, great liberty is often taken in replacing the physically correct laws by largely ad hoc rules, more or less loosely inspired by the real thing. In part because of this great simplification at the microscopic level, what these models do manage to capture is the wide separation of scales often inherent in the natural systems or phenomena under consideration. Such models should thus be considered as complementary to conventional approaches based on rigorous ab initio formulation of microscopic laws, which often end up severely limited in the range of scales they can capture.

This apology for simple models is also motivated, albeit indirectly, by my pledge not to formally define complexity. Instead, *you* will have to develop your own intuitive understanding of it, and if along the way you come up with your own convincing formal definition of complexity, all the better! To pick up on the quote opening this introductory chapter, there is all the difference in the world between theory and practice, between knowledge and know-how. This takes us to the final keyword of this book's title: *handbook*. This is a "how-to" book; its practical aim is to provide material and guidance to allow you to learn about complexity through hands-on experimentation with complex systems. This will mean coding and running computer programs, and analyzing and plotting their output.

1.6 About the Computer Programs Listed in This Book

My favorite book on magnetohydrodynamics opens its preface with the statement, "Prefaces are rarely inspiring and, one suspects, seldom read." I very much suspect so as well, and consequently opted to close this introductory chapter with what would conventionally be preface material, to increase the probability that it will actually be read, because it is really important stuff.

If this book is to be a useful learning tool, it is *essential* for the reader to code up and run programs, and modify them to carry out at least some of the additional exercises and computational explorations proposed at the end of each chapter, including at least a few of the Grand Challenges. Having for many years taught introductory computational physics to the first-semester physics cohort at my home institution, I realize full well that this can be quite a tall order for those without prior programming experience, and, at first, a major obstacle to learning. Accordingly, in developing the models and computer codes listed throughout this book, I have opted to retain the same design principle as in the aforementioned introductory class:

1. There are no programming prerequisites; detailed explanations accompany every computer code listed.

2. The code listings for all models introduced in every chapter must fit on one page, sometimes including basic graphics commands (a single exception to this rule does occur, in chapter 10).

3. All computer programs listed use only the most basic coding elements, common to all computing languages: arrays, loops, conditional statements, and functions. Appendix A provides a description of these basic coding elements and their syntax.

4. Computing-language-specific capabilities, including predefined high-level functions, are avoided to the largest extent possible.

5. Clarity and ease of understanding of the codes themselves is given precedence over run-time performance or "coding elegance."

Each chapter provides a complete code listing (including minimal plotting/graphics commands) allowing simulation results presented therein to be reproduced. These are provided in the programming language Python, even though most of the simulation codes introduced throughout this book were originally designed in the C or IDL programming languages. The use of Python is motivated primarily by (1) its availability as free-of-charge, public-domain software, with excellent on-line documentation, (2) the availability of outstanding public-domain plotting and graphics libraries, and (3) its rising "standard" status for university-level teaching. Regarding this latter point, by now I am an old enough monkey to have seen many such pedagogical computing languages rise and fall (how many out there remember BASIC? APL? PASCAL?). However, in view of the third design principle above, the choice of a computing language should be largely irrelevant, as the source codes[4] should be easy to "translate" into any other computing language. This wishful expectation was subjected to a real-life reverse test in the summer of 2015: two physics undergraduates in my department worked their way through an early, C-version of this book, recoding everything in Python. Both had some prior coding experience in C, but not in Python; nonetheless few difficulties were encountered with the translation process.

The above design principles also have significant drawbacks. The simulation codes are usually very suboptimal from the point of view of run-time speed. Readers with programming

---

[4]Strictly speaking, what I refer to here as "source codes" should be called "scripts," since Python instructions are "interpreted," rather than compiled and executed. While well aware of the distinction, throughout this book I have opted to retain the more familiar descriptor "source code."

experience, or wishing to develop it, will find many hints for more efficient computational implementation in some of the exercises included at the end of each chapter. Moreover, the codes are often not as elegant as they could be from the programming point of view. Experienced programmers will undoubtedly find some have a FORTRAN flavor, but so be it. Likewise, seasoned Python programmers may be shocked by the extremely sparse use of higher-level Python library functions, which in many cases could have greatly shortened the coding and/or increase run-time execution speed. Again, this simply reflects the fact that code portability and clarity have been given precedence.

A more significant, but unfortunately unavoidable, consequence of my self-imposed requirement to keep computational (as well as mathematical and physical) prerequisites to a minimum, is that some fascinating natural complex phenomena had to be excluded from consideration in this book; most notably among these perhaps, is anything related to fluid turbulence or magnetohydrodynamics, but also some specific natural phenomena such as solar flares, geomagnetic substorms, Earth's climate, or the workings of the immune system or the human brain, if we want to think *really* complex. Nonetheless, a reader working diligently through the book and at least some of the suggested computational explorations, should come out well equipped to engage in the study and modeling of these and other fascinating instances of natural complexity.

1.7 Suggested Further Reading

Countless books on complexity have been published in the last quarter century, at all levels of complexity (both mathematically and conceptually speaking!). Among the many available non-mathematical presentations of the topic, the following early best seller still offers a very good and insightful broad introduction to the topic:

Gell-Mann, M., *The Quark and the Jaguar*, W.H. Freeman (1994).

For something at a similar introductory level but covering more recent developments in the field, see, for example,

Mitchell, M., *Complexity: A Guided Tour*, Oxford University Press (2009).

At a more technical level, the following remains a must-read:

Kauffman, S.A., *The Origin of Order*, Oxford University Press (1993).

With regard to natural complexity and the hands-on, computational approach to the topic, I found much inspiration in and learned an awful lot from

Flake, G.W., *The Computational Beauty of Nature*, MIT Press (1998).

Complexity is covered in chapters 15 through 19, but the book is well worth reading cover to cover. In the same vein, the following is a classic not to be missed:

Resnick, M., *Turtles, Termites, and Traffic Jams*, MIT Press (1994).

Statistical physics and thermodynamics is a standard part of the physics cursus. In my department the topic is currently taught using the following textbook:

Reif, F., *Fundamentals of Statistical and Thermal Physics*, reprint, Waveland Press (2009).

For a more modern view of the subject, including aspects related to complexity, I very much like the following book:

Sethna, J. P., *Entropy, Order Parameters, and Complexity*, Oxford University Press (2006)

Good nonmathematical presentations aimed at a broader audience are however far harder to find. Of the few I know, I would recommend chapter 4 in

Gamow, G., *The Great Physicists from Galileo to Einstein* (1961), reprint, Dover (1988).

The literature on chaos and chaotic dynamics is also immense. At the nontechnical level, see, for example,

Gleick, J., *Chaos: Making a New Science*, Viking Books (1987).

For readers fluent in calculus, I would recommend

Mullin, T., ed., *The Nature of Chaos*, Oxford University Press (1993);

Hilborn, R.C., *Chaos and Nonlinear Dynamics*, 2nd ed., Oxford University Press (2000).

The logistic model of population growth is discussed in detail in chapters 5 and 6 of Mullin's book. The functional and structural relationship between chaos and complexity remains a nebulous affair. Those interested in the topic can find food for thought in

Prigogine, I., and Stengers, I., *Order Out of Chaos*, Bantam Books (1984);

Kaneko, K., "Chaos as a source of complexity and diversity in evolution," in *Artificial Life*, ed. C. Langton, MIT Press (1995).

The M.C. Escher foundation maintains a wonderful website, where reproductions of Escher's art can be viewed and enjoyed; see

http://www.mcescher.com/.

Anyone interested in Escher's use of symmetry and transformations should not miss

Schattschneider, D., *Escher: Visions of Symmetry*, 2nd ed., Abrams (2003).

Finally, next time you have a good block of reading time in front of you and are in the mood for a mind-bending journey into complexity in the broadest sense of the word, fasten your seat belts and dive into

Hofstadter, D.R., *Gödel, Escher, Bach*, Basic Books (1979).

2

ITERATED GROWTH

The bewildering array of complex shapes and forms encountered in the natural world, from tiny crystals to living organisms, often results from a growth process driven by the repeated action of simple "rules." In this chapter we examine this general idea in the specific context of *cellular automata* (hereafter, often abbreviated as CAs), which are arguably the simplest type of computer programs conceivable. Yet they can sometimes exhibit behaviors that, by any standard, can only be described as extremely complex.

Cellular automata also exemplify, in a straightforward computational context, a recurring theme that runs through all instances of natural complexity to be encountered in this book: simple rules can produce complex global "patterns" that cannot be inferred or predicted even when complete, a priori knowledge of these rules is at hand.

2.1 Cellular Automata in One Spatial Dimension

Imagine a one-dimensional array of contiguous cells, sequentially numbered by an index j starting at $j = 0$ for the leftmost cell. Each cell can be "painted" either white or black, with the rule for updating the jth cell depending only on its current color and those of the two neighboring cells at positions $j - 1$ and $j + 1$. Consider now the following graphical procedure: At each iteration, the CA looks like a linear array of cells that are either black or white. If we now stack successive snapshots of this row of cells below one another, we obtain a two-dimensional spatiotemporal picture of the CA's evolution, in the form of a (black and white) pixelated image such as formed on the CCD of a digital camera, except that each successive row of pixels captures an iteration of the growth process, rather than the vertical dimension of a truly 2-D image.

The simple question then is, starting from some given initial pattern of white and black cells, how will the array evolve in response to the repeated application of the update rule to every cell of the array? As a first example, consider the following very simple CA update rule:

- **First rule.** Cell j becomes (or stays) black if one or more of the neighbor triad $[j - 1, j, j + 1]$ is black; otherwise it becomes (or stays) white.

The top panel of figure 2.1 shows the first 20 iterations of a CA abiding by this rule, starting from a single black cell in the middle of an otherwise all-white array. On this spatiotemporal diagram, the sideways growth of the CA translates into a black triangular shape expanding by one cell per iteration from the single initial black cell. Now start again from a single black cell but adopt instead the following, equally simple, second update rule:

- **Second rule.** Cell j becomes (or stays) black if either or both of its neighbors $j - 1$ and $j + 1$ are black; otherwise it stays (or becomes) white.

This yields the pattern plotted in the bottom panel of figure 2.1. The global shape is triangular again, but now the interior is a checkerboard pattern of white and black cells alternating regularly in both the spatial and temporal dimensions. With just a bit of thinking, these two patterns could certainly have been expected on the basis of the above two rules.

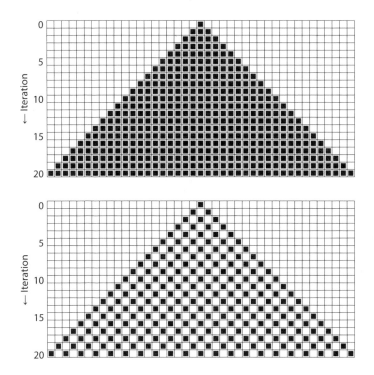

Figure 2.1. The first 20 iterations of a 1-D CA abiding by the first (top) and second (bottom) update rules introduced in the text, in both cases starting from a single black cell at the center of the array (iteration 0, on top). The horizontal direction is the "spatial" dimension of the 1-D CA, and time (iterations) runs downward.

But is it always the case that simple CA rules lead to such simple, predictable spatiotemporal patterns? Consider now the following update rule:

- **Third rule.** Cell j becomes (or stays) black if one and only one of its two neighbors $j - 1$ or $j + 1$ is black; otherwise it stays (or becomes) white.

This is again a pretty simple update rule, certainly as simple as our first and second rules. The top panel of figure 2.2 shows the pattern resulting from the application of this rule to the same initial condition as before, namely, a single black cell at the array center. The globally triangular shape of the structure is again there, but the pattern materializing within the structure is no longer so simple. Many white cells remain, clustered in inverted triangles of varying sizes but organized in an ordered fashion. This occurs because our third rule implies that once the cells have reached an alternating pattern of white/black across the full width of the growing structure, as on iterations 3, 7, and 15 here, the rule forces the CA to revert to all white at the next iteration, leaving only two black cells at its right and left extremities. Right/left symmetrical growth of new black cells then resumes from these points, replicating at each end the triangular fanning pattern produced initially from the single original black cell. Growth thus proceeds as a sequence of successive branching, fanning out, and extinction in the interior.

The bottom panel of figure 2.2 displays 512 iterations of the same CA as in the top panel, with the cell boundaries now omitted. Comparing the top and bottom panels highlights the fundamental difference between the "microscopic" and "macroscopic" views of the structure. On the scale at which the CA is operating, namely, the triad of neighboring cells, a cell is either white or black. On this microscopic scale the more conspicuous pattern to be noticed in the top panel of figure 2.2 is that blacks cells always have white cells for neighbors on the right, left, above, and below, but no so such "checkerboard" constraint applies to white cells (unlike in the bottom panel of figure 2.1). At the macroscopic level, on the other hand, the immediate perception is one of recursively nested white triangles. In fact, the macroscopic triangular structure can be considered as being made from three scaled-down copies of itself, touching at their vertices; each of these three copies, in turn, is made up of three scaled-down

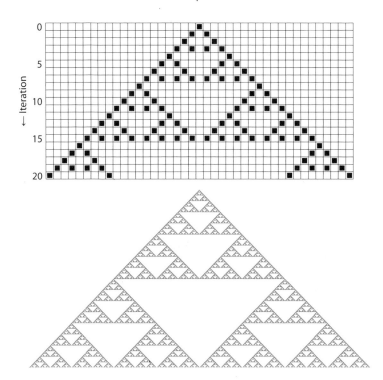

Figure 2.2. The top panel is identical in format to figure 2.1, but now shows the structure produced after 20 iterations by the third CA update rule, introduced in the text. The bottom panel shows the same CA, now pushed to 512 iterations, with cell boundaries removed for clarity.

copies of itself, and so on down to the "microscopic" scale of the individual cells.[1]

[1] This structure belongs to a class of geometrical objects known as *Sierpinski triangles*. It can be constructed by a number of alternative geometrical procedures. The simplest consists in drawing a first triangle, then partitioning it into four smaller triangles by tracing three straight-line segments joining the edge centers, then repeating this process for the three outer triangles so produced, and so on. The CA, in contrast, generates the same macroscopic structure via a directional iterative spatiotemporal growth process.

This type of recursive nesting is a hallmark of *self-similarity*, and flags the structure as a *fractal*. For now you may just think of this concept as capturing the fact that successive zooms on a small part of a structure always reveal the same geometrical pattern, like with the bifurcation diagram for the logistic map encountered in chapter 1 (see figure 1.2). More generally, self-similarity is a characteristic feature of many complex systems, and will be encountered again and again throughout this book.

Consider, finally, a fourth CA update rule, hardly more complicated than our third:

- **Fourth rule.** Cell j becomes (or stays) black if one and only one of the triad $[j - 1, j, j + 1]$ is black, or if only j and $j + 1$ are black; otherwise it stays (or becomes) white.

This rule differs from the previous three in that it now embodies a directional bias, being asymmetrical with respect to the central cell j: a white cell at $j - 1$ and black cells at j and $j + 1$ will leave j black at the next iteration, but the mirror configuration, $j - 1$ and j black and $j + 1$ white, will turn j white. The top panel of figure 2.3 shows the first 20 iterations of this CA, as usual (by now) starting from a single black cell. The expected symmetrical triangular shape is there again, but now the interior pattern lacks mirror symmetry, not surprisingly so perhaps, considering that our fourth rule itself lacks right/left symmetry. But there is more to it than that. Upon closer examination one also realizes that the left side of the structure shows some regularities, whereas the right half appears not to. This impression is spectacularly confirmed upon pushing the CA to a much larger number of iterations (bottom panel). On the right the pattern appears globally random. As with our third rule, inverted white triangles of varying sizes are generated in the course of the evolution, but their spatial distribution is quite irregular and does not abide by any obvious recursive nesting pattern. On the left,

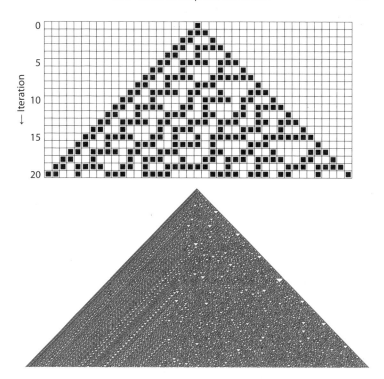

Figure 2.3. Identical in format to figure 2.2, but now for the fourth CA update rule introduced in the text.

in contrast, the pattern is far more regular, with well-defined structures of varying periodicities recurring along diagonal lines running parallel to the left boundary of the structure.

With eight possible three-cell permutations of two possible states (white/black, or 0/1, or whatever) and evolution rules based on three contiguous cells (the cell itself plus its right and left neighbors), there exist 256 possible distinct evolutionary rules.[2]

---

[2]Describing CA rules in words, as we have done so far, can rapidly become quite awkward. A much superior and compact description can be made using an 8-bit binary string, with each bit giving the update (black = 0 and white = 1,

Even if always starting from a single active cell, as in figure 2.1, these various rules lead to a staggering array of patterns, going from triangular wedges, repeating checkerboard or stripe patterns, simple or not-so-simple patterns propagating at various angles, nested patterns (as in figure 2.2), mixtures of order and disorder (as in figure 2.3), all the way to complete randomness. You get to explore some of these in one of the computational exercises proposed at the end of this chapter.

These 256 1-D CA rules can be divided fairly unambiguously into four classes, according to the general properties of the end state they lead to, that are independent of the initial condition.[3]

- Class I CAs evolve to a stationary state; our first rule (figure 2.1) offers an example, although keep in mind that the stationary state need not be all black or all white.
- Class II CAs evolve into a periodic configuration, where each cell repeatedly cycles through the same set of states (which may differ from one cell to the next). Our second rule is a particularly simple exemplar of this class.
- Class III CAs evolve into a nonperiodic configuration. Even though figure 2.2 looks quite regular, our third rule belongs, in fact, to this class, as you will get to verify in one of the computational exercises suggested at the end of this chapter.

---

say) associated with one of the eight possible permutations of black/white on three cells. As a bonus, interpreting each such string as the binary coding of an integer yields a number ranging from 0 to 255, which then uniquely labels each possible rule. Chapter 3 of the book by Wolfram, cited in the bibliography, describes this procedure in detail. Under this numbering scheme, the four rules introduced above are numbered 254, 250, 90, and 30, respectively.

[3]The classification is best established through the use of a random initial condition, where every cell in the initial state is randomly assigned white or black with equal probability. In such a situation, it is also necessary to introduce periodic boundary conditions, as if the 1-D CA were in fact defined on a closed ring: the last, rightmost cell acts as the left neighbor of the first, leftmost cell, and vice versa.

- Class IV CAs collect everything else that does not fit into the first three classes; they are also, in some sense, the more interesting rules in that they yield configurations that are neither stationary, periodic, nor completely aperiodic.

Figure 2.4 gives a minimal source code in the Python programming language for running the 1-D two-state CA of this section, with a value of 0 for white cells and 1 for black cells. The CA evolution is stored in the 2-D array image, the first dimension corresponding to time/iteration, and the second to the spatial extent of the CA (line 8). This code uses a single black cell at the lattice center for the initial condition (line 9), and is set up to run the third rule introduced above. The condition "one and only one of the nearest neighbors is black" is evaluated by summing the corresponding numerical values of the cells (line 14); if the sum is 1, then node j turns black at the next iteration (value 1 in array image, line 15), and otherwise remains white (initialized value 0). Periodicity is enforced in the spatial direction (lines 18–19; see appendix D.2 for further detail on this). Upon completion of the CA's evolution over the preset number of iterations, the structure produced is displayed using the imshow() function from the matplotlib.pyplot graphics library (lines 22–23).

2.2 Cellular Automata in Two Spatial Dimensions

Cellular automata are readily generalized to two (or more) spatial dimensions, but the various possible lattice geometries open yet another dimension (figuratively speaking!) to the specification of the CA and its update rules. It will prove useful to adopt an alternative, but entirely equivalent, formulation of CAs based on a *lattice* of interconnected *nodes*, conceptually equivalent to the center of cells in figures 2.1–2.3. Figure 2.5 illustrates the idea, for different types of 2-D lattices using different *connectivities* between neighboring nodes.

```
1  # 1D 2-STATES CELLULAR AUTOMATON
2  import numpy as np
3  import matplotlib.pyplot as plt
4  #----------------------------------------------------------------------
5  N=129                                      # Size of 1D CA
6  n_iter=64                                   # Number of iterations
7  #----------------------------------------------------------------------
8  image=np.zeros([n_iter,N],dtype='int')      # Initialize lattice to white
9  image[0,N/2]=1                               # But set central node to black
10
11 for iterate in range(1,n_iter):             # Iteration loop
12
13     for j in range(1,N-1):                   # Lattice loop
14         if image[iterate-1,j+1]+image[iterate-1,j-1] == 1:  # Third rule
15             image[iterate,j]=1               # Turn node black
16     # End of lattice loop
17
18     image[iterate,0]=image[iterate,N-2]      # Enforce periodicity
19     image[iterate,N-1]=image[iterate,1]
20 # End of iteration loop
21
22 plt.imshow(image,interpolation="nearest")    # Display structure
23 plt.show()
24 # END
```

Figure 2.4. A minimal source code in the Python programming language for the 1-D CA introduced in this section. As with all Python codes given in this book, the Python NumPy library is used. The NumPy function np.zeros() generates an array of whatever size is given to it as an argument and fills it with zeros; here it generates a 2-D array of size n_iter×N, through the argument [n_iter,N]. The code given here operates according to the third CA rule introduced in the text.

On Cartesian lattices in two spatial dimensions (panels A and B), connectivity typically involves either only the 4 nearest neighbors (in red) on the top/bottom/right/left of a given node (in black), or additionally the 4 neighbors along the two diagonals (panel B).[4] Anisotropic connectivities, as in panel C, can be

---

[4]The top/down/right/left 4-neighbor connectivity on a Cartesian lattice is sometimes referred to as the *von Neumann neighborhood*, and the 8-neighbor connectivity as the *Moore neighborhood*. See appendix D.1 for more on these matters.

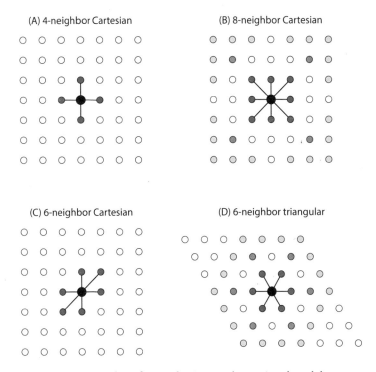

Figure 2.5. Examples of some lattices and associated nodal connectivities, as indicated by the line segments connecting the central black node to its nearest neighbors in red. The meaning of the orange- and yellow-colored nodes in panels B and D will be elucidated in the text.

reinterpreted as changes in lattice geometry. Upon introducing a horizontal displacement of half an internodal distance per row and compressing vertically by a factor $\sin(\pi/3) \simeq 0.866$, as shown in panel D, one obtains a regular triangular lattice with 6-neighbor connectivity. From the point of view of CA evolutionary rules, the two lattices in panels C and D are topologically and operationally equivalent. What is interesting in practice is that whether triangular or Cartesian, these lattices can all be conveniently stored as 2-D arrays in the computer's memory, and the "true" geometry becomes set by the assumed connectivity.

We first restrict ourselves to the following very simple 2-D CA rule:

- A node becomes active if one and only one of its neighboring nodes is also active.

Note that such a rule has no directional bias other than that imposed by the lattice geometry and connectivity: any one active node will do. However, a noteworthy difference from the 1-D CA rules considered previously is that here, once activated a node remains activated throughout the remainder of the iterative process. Nonetheless, as far as rules go, this is probably about as simple as it could get in this context. Figure 2.6 lists a minimalistic Python source code for this automaton, defined on a triangular lattice with 6-neighbor connectivity. Note the following:

1. The code is structured as an outer temporal loop running a preset number of temporal iterations `n_iter` (lines 16–28), inside which two nested loops over each lattice dimension (lines 20–21) carry out the activation test.

2. The connectivity is enforced through the use of the 1-D template arrays `dx` and `dy` in which are hardwired the relative locations, measured in lattice increments, of the connected neighbors (lines 9–10); these 1-D arrays are then used to access the 2-D array `image` which stores the state of the CA proper (lines 23–24).

3. If a cell is to become active at the next iteration, its new state is temporarily stored in the 2-D work array `update` (line 26), which is reset to zero at the beginning of each temporal iteration (line 18); only once all nodes have been tested is the lattice array

```
 1  # 2D 2-STATES CELULAR AUTOMATON ON TRIANGULAR LATTICE
 2  import numpy as np
 3  import matplotlib.pyplot as plt
 4  #----------------------------------------------------------------------
 5  N=24                                      # Size of 2D CA
 6  n_iter=10                                 # Number of iterations
 7  n_neighbor=6                              # Number of connected neighbors
 8  #----------------------------------------------------------------------
 9  dx=np.array([ 1, 0,1,-1,0,-1])           # nearest neighbor template
10  dy=np.array([-1,-1,0, 1,1, 0])
11  image=np.zeros([N,N],dtype='int')        # Initialize lattice to white...
12  image[N//2,N//2]=1                        # ...except central node to black
13  plt.scatter(N//2,N//2)                    # Set up plot, with central node
14  plt.axis([0,N,0,N])
15  plt.axes().set_aspect('equal')
16  for iteration in range(1,n_iter):         # Iteration loop
17
18      update=np.zeros([N,N],dtype='int')    # Set/reset evolution array
19
20      for i in range(1,N-1):                # Lattice loops
21          for j in range(1,N-1):
22              cumul=0
23              for k in range(0,n_neighbor):  # Loop over nearest-neighbor
24                  cumul+=image[i+dx[k],j+dy[k]]
25              if image[i,j]==0 and cumul==1:  # Only one active neighbor
26                  update[i,j]=1              # Activate node
27                  plt.scatter(j+(i-N//2)/2.,N//2+0.866*(i-N//2)) # Plot node
28      # End of lattice loops
29
30      image+=update                         # Synchronous update of CA
31
32  # End of iteration loop
33  plt.show()                                # Display structure
34  # END
```

Figure 2.6. Source code in the Python programming language for a 2-D CA defined over a triangular lattice with 6-neighbor connectivity. This is a minimal implementation, emphasizing conceptual clarity over programming elegance, code length, or run-time speed. The various matplotlib instructions (lines 13–15, 27, and 33) display the final structure, with activated nodes shifted horizontally and compressed vertically (line 27) so as to correctly display the structure in physical space, as in figure 2.7.

image updated (line 30). This *synchronous update* is necessary, otherwise the lattice update would depend on the order in which nodes are tested in the first set of lattice loops, thus introducing an undesirable spatial bias that would distort growth.

Figure 2.7. Structure growth generated by the 2-D CA with the simple one-neighbor rule on a triangular, 6-neighbor lattice. The top-left image shows the structure after 5 iterations, and the other images display the subsequent evolution at a 3-iteration cadence, the growth sequence being obvious. In the notation of section 2.3, this rule is written as $(1) + 1$.

Now consider growth beginning from a single occupied node (the "seed") at the center of a triangular lattice. The first three steps of the iterated growth process are illustrated via the color coding of lattice nodes in figure 2.5D. Starting from a single active node, the next iteration is a hexagonal ring of 6 active nodes (in red) surrounding the original active node (in black). At the next iteration, only the 6 nodes colored in orange abide by the one-neighbor-only activation rule, but at the following iteration, each of these 6 "branches" will generate an arc-shaped clump of 3 active nodes (in yellow) at its extremity. Figure 2.7 picks up the growth at iteration 5 (top left), with subsequent frames plotted at a cadence of 3 iterations. As the six "spines" grow radially outward, the faces of the growing structure eventually fill inward from the corners, until a hexagonal shape is produced; from that point on, growth can pick up again only at the six corners, and later toward the centers of the faces, eventually adding another "layer" to the growing structure. The broken, concentric white hexagons within the structure, plotted in the bottom-right corner of figure 2.7, are the imprint of this layered growth process. Evidently, here the sixfold symmetry of the connectivity remains reflected in the global, "macroscopic" shape of the growing structure; this could perhaps have been expected, but certainly not the intricacies of details produced within the structure itself. In fact there is much more to these details than meets the eye; step back a bit to view the bottom-right structure from a distance, and it will hardly be distinguishable from the middle-left structure viewed at normal reading distance. This is again an indication of self-similarity.

All this being said, looking at figure 2.7, the first thing that comes to mind is, of course, snowflakes! It might appear ludicrous to suggest that our very artificial computational setup—triangular lattice, neighbor-based growth rule, etc.—has anything to do with the "natural" growth of snowflakes, but we will have occasion to revisit this issue in due time.

A similar 2-D CA simulation can be run on a Cartesian lattice with 8-neighbor connectivity, starting again with a single active node at the lattice center. All that is needed is to append two elements to the stencil arrays dx and dy in the code listed in figure 2.6. The first four steps of the growth process are again indicated by the nodal color coding in figure 2.5B. The first iteration produces a 3×3 block of active nodes, but at the next iteration our one-neighbor rule makes growth possible only along diagonals quartering this 3 × 3 block (orange nodes). The next iteration (yellow nodes) generates a 5-node 90-degree wedge about each of the four extrusions generated at the preceding iteration; except for the fourfold symmetry, this is essentially the same growth pattern observed in sixfold symmetry on the triangular lattice (figure 2.5D). Figure 2.8 picks up the growth process at iteration 5, and subsequent frames are plotted at a 3-iteration cadence, as in figure 2.7. Growth now proceeds from the corners of the squares, which spawn more squares at their corners, and so on as the structure keeps growing, once again in a self-similar fashion.

2.3 A Zoo of 2-D Structures from Simple Rules

We henceforth restrict ourselves to a Cartesian lattice with 8-neighbor connectivity, and introduce a generalized 8-neighbor activation rule as follows: a node becomes active if either n_1 or n_2 neighboring nodes are active. We also include in the rule the number s of "seed" active nodes used to initialize the growth process. We write all this as

$$(n_1, n_2) + s, \quad 1 \leq n_1, n_2 \leq 8, \quad n_1 < n_2. \qquad (2.1)$$

A specific example will likely help more than further explanation: The rule $(1, 5) + 1$ means that we start from one active node ($s = 1$). A node becomes active if either 1 or 5 neighboring nodes are active, and remains inactive otherwise, namely, if it has either 0, 2, 3, 4, 6, 7, or 8 active neighbors. Once again, no directional bias is introduced, as it doesn't matter where the 5 active nodes (say) are located in the 8-node group of neighboring nodes. Under this notation, the rules used to grow the structure in figure 2.7 would be written as $(1) + 1$.

Figure 2.9 shows a sample of structures grown using various such rules, as labeled. The variety of structures produced even by this narrow subset of rules is quite staggering, including again

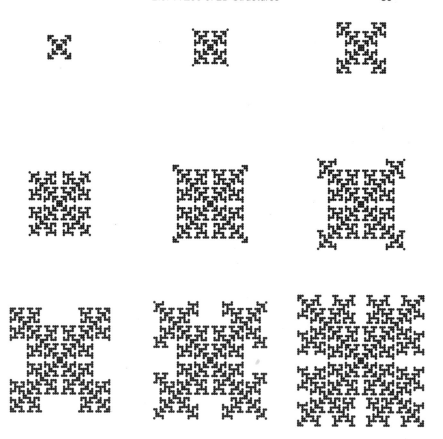

Figure 2.8. Structure generated by the 2-D CA with the simple rule (1) + 1, now on a Cartesian 8-neighbor lattice. The top-left image shows the structure after 5 iterations, and the other images display the subsequent evolution at a 3-iteration cadence, the growth sequence again being obvious.

self-similarity, a mixture of order and disorder, compact structures porous or solid, etc. Some rules, such as (3, 6) + 5, do not even generate regular outward growth, as extrusions fold back inward to fill deep crevices left open in earlier phases of the iterated growth process.

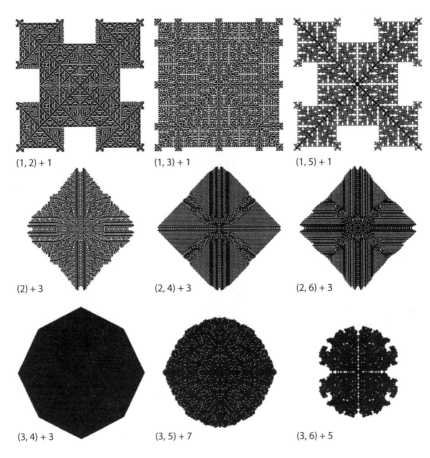

(1, 2) + 1   (1, 3) + 1   (1, 5) + 1

(2) + 3   (2, 4) + 3   (2, 6) + 3

(3, 4) + 3   (3, 5) + 7   (3, 6) + 5

Figure 2.9. A zoo of structures grown by the 2-D CA on a Cartesian 8-neighbor lattice operating under a variety of rules, as labeled. All automata were executed for 100 iterations, except for the bottom three, for which 200 iterations were executed.

Staring at these and other structures generated by other specific incarnations of the two-member rule (2.1), one is naturally tempted to extract some general trends. For example, rules beginning with "1" produce squares that grow by spawning more squares at their corners, in a manner qualitatively similar to the basic $(1) + 1$ rule; the numerical choice for n_2 affects primarily the internal pattern. Rules beginning with a "2" on the other hand, produce diamond-shaped structures, with ordered and disordered internal regions, growing along their four approximately linear faces; the numerical choice for n_2 mostly affects the relative importance of ordered and disordered regions in the interior. Rules with a "3" produce compact structures, sometimes solid sometimes porous, with various patterns of symmetry about vertical, horizontal, or diagonal axes present at the global scale. Now, if you find this convincing on the basis of figure 2.9, try running a simulation for rule $(3, 7) + 5$ and reconsider your position!

The overall conclusion of our relatively limited explorations of 2-D CAs remains the same as with the 1-D CAs considered previously: simple, microscopic growth rules can produce macroscopic structures ranging from highly regular to highly "complex," and very, very few of these structures could have been anticipated knowing only the lattice geometry and the growth rules.

2.4 Agents, Ants, and Highways

In the "classical" CAs considered thus far, the active elements are the lattice nodes themselves, and so they are fixed in space by definition. Another mechanism for iterated growth involves active elements moving on and interacting with the lattice (and/or with each other), according once again to set rules. Henceforth, such active elements will be defined as "agents." For example, an agent known as an "ant" moves and operates on a lattice as follows, from one iteration to the next:

- Move forward.
- If standing on a white node, paint it black and turn right by 90 degrees.
- If standing on a black node, paint it a whiter shade of pale (white) and turn left by 90 degrees.

```
 1  # HIGHWAY BUILDING BY LANGTON'S ANT
 2  import numpy as np
 3  import matplotlib.pyplot as plt
 4  #----------------------------------------------------------------------
 5  N     =300                         # Lattice size
 6  n_iter=20000                       # Number of temporal iterations
 7  #----------------------------------------------------------------------
 8  x_step=np.array([0,-1,0,1])        # Template arrays for steps
 9  y_step=np.array([1,0,-1,0])
10  image=np.zeros([N,N],dtype='int')  # Initialize lattice array, all white
11  ix=N//4                            # Ant's starting position in x
12  iy=N//4                            # Ant's starting position in y
13  direction=1                        # Ant's starting direction, North
14
15  for iteration in range(0,n_iter):  # Temporal loop
16
17      ix+=x_step[direction]          # Ant moves
18      iy+=y_step[direction]
19      ix=(N+ix) % N                  # Enforce periodicity in x
20      iy=(N+iy) % N                  # Enforce periodicity in y
21
22      if image[iy,ix] == 0:          # On a white node
23          update=1                   # Paint it black...
24          direction+=1               # ...and turn right...
25          direction=direction % 4    # ...but stay within step array
26      else:                          # On a black node
27          update=-1                  # Paint it white...
28          direction-=1               # ...and turn left...
29          direction=(4+direction) % 4 # ...but stay within step array
30
31      image[iy,ix]+=update
32  # End of temporal loop
33  plt.imshow(image,interpolation="nearest")
34  plt.show()                         # Display final structure
35  # END
```

Figure 2.10. Source code in the Python programming language for an "ant" agent abiding by the rules introduced in section 2.4. This code generates the simulation plotted in the top panel of figure 2.11 (minus the colored dots and insets).

These are pretty simple behavioral rules, yet they hold surprises in store for us.

Figure 2.10 gives a simple numerical implementation of these behavioral rules. As with most codes listed throughout this book, logical clarity and readability have been given precedence over programming elegance, code length, or run-time speed, and computing-language-specific capabilities are systematically avoided. Note the following:

1. The code is again structured as an outer temporal loop running over a preset number of temporal iterations n_iter (starting on line 15).

2. The two arrays x_step and y_step store the x- and y-increments associated with the four possible displacements, in the order down, left, up, right (lines 8–9). These are used to update the ant's position on the lattice (ix, iy) as per the ant's direction, stored in the variable direction. Under this ordering convention, turning right requires incrementing direction by $+1$ (line 24), and turning left by -1 (line 28).

3. The modulus operator % is used to enforce periodicity for the ant's position (lines 19–20) and stepping direction (lines 25 and 29). The instruction $a\%b$ returns the remainder of the division of a by b, for example $7\%3 = 1$, $2\%3 = 2$, $3\%3 = 0$. See appendix A.3 for more on the use of the modulus operator in Python.

4. The *change* in the lattice state at the ant's position is first calculated (variable update) and the lattice is updated (line 31) only once the **if. . . else** construct is exited. This is needed because the lattice state at the ant's position sets the operating condition of this logical structure, so changing its value *within* its blocks of instructions is definitely not a good idea in most programming languages.

The top panel in figure 2.11 shows the structure built by a single ant moving on a 300×300 lattice, starting at the location marked by the red dot, and initially pointing north (top of the page). The initial state of the lattice is all white nodes. These are the parameter settings and initial conditions implemented in

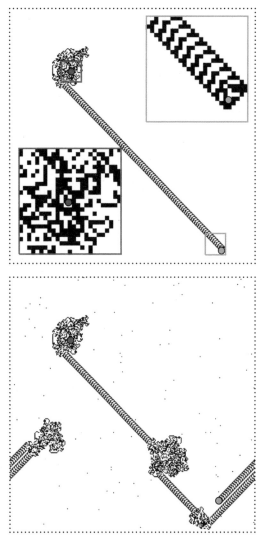

Figure 2.11. Highway building by an "ant agent" in a clean (top; 20,000 iterations) and noisy (bottom; 51,000 iterations) background environment. The solid dots show the starting (red) and ending (green) position of the ant, with the insets in the top panel providing close-ups of the lattice about these two points. The lattice is assumed periodic both horizontally and vertically. See appendix D.2 for more on periodic boundary conditions on lattices.

the code listed in figure 2.10. The first few thousands iterations, shown in the inset framed in red, produce, if not strictly random, at least a highly disordered clump of white and black nodes. But after a bit more than 10,000 time steps, a switch to a different behavior takes place. The ant now executes a periodic series of steps, involving a lot of backtracking but also a net drift velocity along a diagonal with respect to the lines of the Cartesian lattice, leaving behind in its trail a highly ordered, spatially periodic pattern of white and black nodes (see green inset). This behavior has been dubbed "highway building," and it could hardly have been expected on the basis of the ant's simple behavioral rules. Highway building always proceeds along 45-degree diagonals, and once started would go on forever on an infinite-sized lattice. The fact that the highway points to the southeast in figure 2.11 is determined by the initial conditions: all white nodes and the ant pointing north.

In practice, simulations such as in figure 2.11 are carried out on a finite-sized lattice, on which horizontal and vertical periodicity is enforced. So here, pushing the simulation further would eventually lead to the ant (and its highway) leaving the lattice near the southeast corner, to reappear near the northwest corner, still heading southeast, eventually hitting the structure it just built. This throws the ant into a fit; disordered (re)painting prevails for a while, forming a structure statistically similar to that characterizing the first 10^4 iterations, but after many thousands of iterations highway building resumes, in a direction orthogonal to that of the original highway, to the southwest here. As the lattice fills with blotches of disorder and stretches of highway crossing each other, highway building becomes increasingly difficult, and if the evolution is pushed sufficiently far, on any finite-sized lattice the end result is randomness.

Highway building is a pretty delicate process that is easily disturbed. The bottom panel of figure 2.11 shows what happens when a small number of randomly selected lattice nodes are painted black before the ant starts moving. At first, the evolution proceeds as before, and highway building toward the southeast begins, but soon the ant hits one of the randomly distributed black nodes, triggering disordered painting. Highway building eventually resumes, still toward the southeast, until another black node is encountered, triggering another, shorter, disordered episode that ends with highway building resuming now toward the northeast; and so on over the 51,000 iterations over which this specific simulation was pursued.

2.5 Emergent Structures and Behaviors

We have barely touched the realm of structures and behaviors that can be produced by CAs and CA-like systems. Nonetheless, the take-home message of this chapter should be already clear: very simple rules can produce very complex-looking structures. But should we really be calling these structures "complex" if their generating rules are simple? Or do we remain tied to an intuitive definition of "complex" relying on our visual perception of structures and behaviors? Students of complexity have been rattling their brains over that one for quite a while now.

If defining complexity is hard, coming up with an unambiguous and universal quantitative *measure* of complexity is perhaps the next hardest thing. Many such measures have been defined, and can be categorized in terms of the question they attempt to answer; given a complex structure or system, the following three queries are all pertinent:

1. How hard is it to describe?
2. How hard is it to create?
3. How is it dynamically organized?

Consider, for example, the measure known as *algorithmic complexity*, namely, the length of the smallest computer program that can generate a given output—a spatial pattern, a time series, a network, whatever. It may appear eminently reasonable to suppose that more complex patterns require longer programs; simulating the evolving climate certainly requires much longer code (and a lot more computer time!) than simulating the harmonic oscillation of a frictionless pendulum. It seems to make sense, but we need to look no further than the simple 1-D CAs investigated in section 2.1 to realize the limitations of this measure of complexity. The CAs of figures 2.1, 2.2, and 2.3 can be produced by programs of exactly the same length, yet they could hardly be considered "equally complex."

Our brief foray into cellular automata also highlights a theme that will recur throughout this book and that is almost universally considered a defining feature of natural complexity: *emergence*. This term is used to refer to the fact that global structures or behaviors on macroscopic scales cannot be reduced to (or inferred from) the rules operating at the microscopic level of individual components making up the system; instead, they emerge from the *interactions* between these components. Synthetic snowflakes and ant highways are such examples of emergence, and are by no means the last to be encountered in this book.

2.6 Exercises and Further Computational Explorations

1. Use the 1-D CA code of figure 2.4 to explore the behavior of the four CA rules introduced in section 2.1 when starting from a random initial condition, i.e., each cell is randomly assigned to be black or white with equal probability. If needed, see appendix C.2 for a quick start on generating uniformly distributed random deviates in Python. Make sure also to enforce periodic boundary conditions (see appendix D.2).

2. The aim of this exercise is to explore further the patterns produced by 1-D CA rules, all of which are relatively easy to implement in the source code of figure 2.4. The following five individual rules produce patterns qualitatively distinct from those already

examined in section 2.1. Unless the pattern looks really trivial, make sure to run the CA for enough iterations to ascertain its long-term behavior.

- Cell j becomes (or stays) black only if both $j - 1$ and j are white; otherwise the cell stays (or becomes) white.
- Cell j becomes (or stays) black if any two cells of the triad $[j - 1, j, j + 1]$ are black, or if both $j - 1$ and j are white; otherwise the cell stays (or becomes) white.
- Cell j becomes (or stays) white if $j - 1$ and j are both black, or if the triad $[j - 1, j, j + 1]$ are all white; otherwise the cell stays (or becomes) black.
- Cell j becomes (or stays) black if cell $j - 1$ and at least one of the pair $[j, j + 1]$ are black, or if the triad $[j - 1, j, j + 1]$ are all white; otherwise the cell stays (or becomes) white.
- Cell j becomes (or stays) white if $j + 1$ and j are both white, or if the triad $[j - 1, j, j + 1]$ are all black; otherwise the cell stays (or becomes) black.

The last two rules, in particular, should be iterated over many hundreds of iterations over a large lattice; the patterns produced are particularly intriguing. You should also run these five rules starting from a random initial condition. To which CA class does each belong?

3. Modify the code of figure 2.6 to introduce two-member rules such as those used to produce figure 2.9 on the 6-neighbor triangular lattice. Explore the growth produced by the set of rules $(1, 2) + 1$ through

$(1, 6) + 1$. Is lattice structure always imprinted on global shape?

4. Modify the code of figure 2.6 to operate on a Cartesian 8-neighbor lattice, and explore the sensitivity to initial conditions for rules $(3, 4) + n$ and $(3, 5) + n$. More specifically, consider the growth produced by using either $n = 3$, 4, or 5 active nodes, organized either linearly, as a 2×2 block, as a diamond-shaped 5-node block, etc. Is the geometry of the initial condition imprinted on the global shape?

5. The Game of Life is one of the most intensely studied 2-D CAs. It is defined on a 2-D Cartesian lattice, periodic horizontally and vertically, with 8-neighbor connectivity. Each lattice node can be in one of two possible states, say "inactive" and "active" (or 0 and 1; or white and black; or dead and alive; whatever), and evolves from one iteration to the next according to the following rules:

   - If an active node has less than two active neighbors, it becomes inactive.
   - If an active node has more than three active neighbors, it becomes inactive.
   - If an inactive node has three active neighbors, it becomes active.
   - If a node has two active neighbors, it remains in its current state.

   This automaton can generate "organisms," i.e., shape-preserving structures moving on the lattice, in some cases interacting with one another or with their environment to produce even more intricate behaviors. Modify the code of figure 2.6 to incorporate the above rules, and run simulations starting from a random initialization of the lattice in which each node is

assigned active or inactive status with equal
probability.

6. The Grand Challenge for this chapter is to explore the
behavior of another interesting ant-like agent, known as
the "termite." Termites move randomly on a lattice on
which "wood chips" (i.e., black) have been randomly
dispersed. The termite's behavioral rules are the
following:

- Random walk until a wood chip is encountered.
- **if** a wood chip is currently being carried, drop
  it at the current position (i.e., next to the one
  just bumped into), and resume the random
  walk.
- **else** pick up the chip bumped into, and resume
  the random walk.

Code this up, perhaps starting from the "ant" code of
figure 2.10. Appendix D.3 may prove useful if you need
a kick-start on how to code up a random walk on a
lattice. How is the distribution of wood chips evolving
with time? Does this change if you let more than one
termite loose on the lattice?

2.7 Further Reading

Pretty much anything and everything that could be written on
CAs can be found in

Wolfram, S., *A New Kind of Science*, Wolfram Media (2002).

The material covered in sections 2.1 and 2.2 follows rather closely
parts of chapters 2 and 8 of this massive tome. The Wikipedia
page on CAs includes a good discussion of the history of CA
research, with copious references to the early literature (viewed
March 2015).

For a succinct and engaging introduction to virtual ants and similar computational insects, see

Resnick, M., *Turtles, Termites, and Traffic Jams*, MIT Press (1994);

as well as chapter 16 in

Flake, G.W., *The Computational Beauty of Nature*, MIT Press (1998).

Chapter 15 of this volume also offers a nice introduction to CAs, including the Game of Life. As far as I know, the general notion of an "agent" has been borrowed from economics and introduced into complexity science by John Holland; for more on this concept see

Holland, J.H., *Hidden Order*, Reading: Addison-Wesley (1995).

Some years ago the term *artificial life* was coined to define a category for computational ants, termites, turmites, boids, and other similarly designed computational critters, as well as those appearing in John Conway's Game of Life. The following collection of papers remains a great overview of this computational zoology and its underlying motivations:

Langton, C., ed., *Artificial Life*, MIT Press (1995).

Langton is actually the designer of the ant agent starring in section 2.4. The Wikipedia page on Langton's ant is worth viewing; it also provides examples of extensions to multiple-state ants, as well as a good sample of references to the technical literature:

http://en.wikipedia.org/wiki/Langton%27s_ant (viewed March 2015),

and, not to be missed, a detailed study of the real thing:

Gordon, D.M., *Ant Encounters: Interaction Networks and Colony Behavior*, Princeton University Press (2010).

On measuring complexity, see

Lloyd, S., "Measures of complexity: A nonexhaustive list," *IEEE Control Syst. Mag.*, **21** (2001);

Machta, J., "Natural complexity, computational complexity and depth," *Chaos*, **21**, 037111 (2011).

The three-questions categorization of complexity measures introduced in section 2.5 is taken from the first of these papers, which lists no less than 42 such measures, in a "nonexhaustive list"! The second paper is part of a focus issue of the research journal *Chaos*, devoted to measures of complexity.

3

AGGREGATION

The structures generated by the CAs of the preceding chapter grew according to lattice-based rules that are both artificial and completely deterministic. By contrast, naturally occurring inanimate structures typically grow by accretion of smaller-sized components, in a manner often far more random than deterministic. Interplanetary dust grains grow by accretion of individual molecules, as well as coalescence with other dust grains. Ice crystals and snowflakes grow by accreting individual water molecules, a process often seeded by a dust grain—some having fallen into the atmosphere from interplanetary space! As spectacularly exemplified by snowflakes, randomly driven accretion can sometimes produce structures combining surprisingly high geometric regularity and complexity.

3.1 Diffusion-Limited Aggregation

We focus here on one specific accretion process, known as *diffusion-limited aggregation* (hereafter, often abbreviated as DLA). The idea is quite simple: particles move about randomly, but stick together when they come into contact. Clumps of particles produced in this manner grow further by colliding with other individual particles, or clumps of particles. Over time, one or more aggregates of individual particles will grow. That is to be expected by the very nature of the aggregation process, but the shape of the aggregates so produced turns out to be nothing like one might have expected.

Conceptually, simulating DLA is simple. Diffusion and random walks are the macroscopic and microscopic representations of the same thermodynamically irreversible process. This equivalence is discussed at some length in appendix C.6. A random walk is defined as a succession of steps taken in directions that vary randomly from one step to another, in a manner entirely independent of the orientation of prior steps. All that is needed to simulate a random walk is really a random number generator.[1] Accordingly, in a DLA simulation, M random-walking "particles" are left to do their usual thing, but whenever any two come within some preset interaction distance, they stick together. Computationally, this means checking $M^2/2$ pairwise distances at every temporal iteration. This $\propto M^2$ scaling rapidly makes the calculation computationally prohibitive at large M. Turning to random walks on a lattice (see appendix D.3) neatly bypasses this problem, since all that needs to be done is to check, for each particle, its nearest-neighbor nodes on the lattice for the presence of a "sticky" particle; the pairwise proximity test now scales as $\propto M$.

In the specific implementation of DLA considered in this chapter, one or more fixed "sticky" particles are placed on the lattice, serving as *seeds* for the growth process. Random-walking particles sticking to these seed particles stop moving upon contact, and become sticky themselves.

3.2 Numerical Implementation

The source code listed in figure 3.1 implements the lattice-based approach to DLA just described, again in a manner far from

---

[1]Appendix C.5 provides an introduction to the mathematical description and statistical properties of random walks.

```python
1   # DIFFUSION-LIMITED AGGREGATION ON A CARTESIAN LATTICE
2   import numpy as np
3   import matplotlib.pyplot as plt
4   #------------------------------------------------------------------------
5   N       =128                          # Lattice size
6   max_iter =100000                      # Max number of temporal iterations
7   n_walkers=1000                        # Number of random walkers
8   #------------------------------------------------------------------------
9   x_step=np.array([-1,0,1,0])           # Template arrays for random walk
10  y_step=np.array([0,-1,0,1])
11  dx      =np.array([-1,0,1,0,-1,1,1,-1]) # Template arrays for sticking
12  dy      =np.array([0,-1,0,1,-1,-1,1,1])
13  grid   =np.zeros([N+2,N+2],dtype='int')  # Lattice array
14  x       =np.zeros(n_walkers,dtype='int') # Walker x-coordinate in nodal unit
15  y       =np.zeros(n_walkers,dtype='int') # Walker y-coordinate in nodal unit
16  status=np.ones(n_walkers,dtype='int')    # Walker status array: all mobile
17  for i in range(0,n_walkers):             # Place walkers on lattice
18      x[i]=np.random.random_integers(0,N-1)
19      y[i]=np.random.random_integers(0,N-1)
20      grid[x[i],y[i]]=1
21  grid[N//2,N//2]=2                     # Introduce sticky central node
22
23  iteration,n_glued=0,0                 # Counters
24  while (n_glued < n_walkers) and (iteration < max_iter):
25      for i in range(0,n_walkers):      # Loop over walkers
26          if status[i] == 1:            # This walker is still mobile
27              ii=np.random.choice([0,1,2,3])      # Pick direction
28              x_new=x[i]+x_step[ii]     # New position on lattice
29              y_new=y[i]+y_step[ii]
30              x_new=(N+x_new) % N       # Periodic boundaries in x,y
31              y_new=(N+y_new) % N
32              grid[x_new,y_new]=1       # Update lattice
33              grid[x[i],y[i]]=0
34              x[i],y[i]=x_new,y_new     # Move walker
35              if 2 in grid[x[i]+dx[:],y[i]+dy[:]]: # Check for sticky neighbor
36                  grid[x[i],y[i]]=2     # Assign sticky status to walker
37                  status[i]=2
38                  n_glued+=1
39          # End of work on this mobile walker
40      # End of loop over all walkers
41      # Graphics command displaying positions of glued walkers could go here
42      iteration+=1
43      print("iteration {0}, glued walkers {1}.".format(iteration,n_glued))
44  # End of temporal loop
45  plt.imshow(grid,interpolation="nearest") # display aggregate as pixel image
46  plt.show()
47  # END
```

Figure 3.1. Source code in the Python programming language for DLA. The random walk uses a four-element template, namely, the two arrays x_step and y_step, to identify the relative position of the four neighboring nodes to which a particle can move at the subsequent random-walk step. A random number from the set {0, 1, 2, 3} is then generated to pick a target lattice node for the random move. See appendix D.3 for more on random walks on lattices.

the most computationally efficient, but at least easy to read and understand. The simulation operates in two spatial dimensions on an $N \times N$ Cartesian lattice, but its generalization to three spatial dimensions is straightforward—although the visualization of results is not. In addition to two arrays x[n_walkers] and y[n_walkers] containing respectively the horizontal and vertical coordinates (in lattice units) of each particle, we also introduce a 2-D array grid, which holds values 0 for an empty node and 1 if the node is occupied by one (or more) random-walking particles. Note that this array must be updated every time a particle makes a move. Elements of the grid array will also be assigned the numerical value 2 wherever a node is occupied by an immobilized sticky particle. Note the following:

1. The code is structured as two nested loops: an outer temporal loop (starting on line 24), and an inner loop (starting on line 25) running over the M particles.

2. Although the lattice is of size N×N, the 2-D array grid has dimensions (N+2)×(N+2) (line 13); the rows and columns 0 and N+1 are "ghost nodes" introduced to avoid overflowing array bounds when testing nearest neighbors, without having to introduce a series of specific conditional statements to modify nearest-neighbor definitions for nodes at the edges of the lattice. See appendix D.1 for more on ghost nodes and lattice boundary conditions.

3. Initialization consists in randomly distributing the particles on the lattice, by assigning them horizontal and vertical positions in the integer arrays x[i] and y[i] (lines 18–19). The corresponding position in the 2-D array grid is initialized to 1 (line 20), to flag the node as being occupied by a moving particle, grid having been initialized to 0 beforehand (line 13).

4. An array status assigns a tag to each random-walking particle: 1 if the particle is mobile, and 2 once it is stuck

next to a sticky particle (line 37). The inner loop then checks for sticky neighbors only for particles still mobile (if `status[i]`==1, line 26).

5. The DLA process begins by assigning "sticky" status to the node located at the center of the lattice (line 21).

6. The outer temporal loop repeats until all particles have been aggregated (i.e., `while(n_glued < n_walkers)`), or until a preset maximum number of temporal iterations (`max_iter`) has been reached.

7. The lattice is considered periodic, so that the positions of particles leaving the lattice are reset to the opposite edge (lines 30–31), as with the ant of section 2.4.

8. The test for a sticky neighbor uses two arrays, `dx` and `dy`, each of length 8 and containing a stencil for the relative positions in x and y of the 8 nearest-neighbor nodes (top+down+right+left+4 diagonals; lines 11–12). Here two bits of Python-specific syntax and operators are used, which are not available in all computing languages (line 35): the indexing `dx[:]` means "loop over all elements of `dx`," and the somewhat self-explanatory conditional statement of the type if `2 in dx[:]` means "if the value 2 is found in any element of array `dx`." Note how the elements of `grid` are accessed in this manner here, but through mathematical operations calculating the corresponding nodal positions within the indexing of `grid`.

9. In order to speed up calculations, here two random-walking particles are allowed to occupy the same node, which is unconventional for particles-on-a-lattice simulations.

10. The final aggregate is displayed by passing the array `grid` as the argument to the `imshow` function from the library matplotlib (lines 45–46).

The DLA algorithm of figure 3.1 is very inefficient, in that it spends a lot of time random-walking particles which are very far from the aggregate, particularly early on in the simulation. Much better run-time performance can be obtained by injecting particles one by one, at random positions along the perimeter of a growing circle circumscribing the growing aggregate. One of the suggested computational exercises at the end of this chapter leads you through the design of a faster DLA code based on this idea.

3.3 A Representative Simulation

Figure 3.2 shows a specific example of a 2-D DLA aggregate, grown here from a single "sticky" seed particle located at the center, with 20,000 random-walking particles initially distributed randomly over the computational plane. This is the setup implemented in the code of figure 3.1. The aggregate grows outward from its seed, as expected, but its shape is anything but an amorphous clump. Instead, the aggregate generates a series of outward-projecting branches, themselves spawning more branches, and so on to the edge of the structure. The particles making up the aggregate in figure 3.2 are color coded according to the order in which they were captured by the growing aggregate, as indicated on the right of the figure. Looking carefully, you should be able to see that growth takes place through capture and successive branching, almost always occurring at or near the tips of existing branches. Unlike with the structures encountered in the preceding chapters, which grew according to deterministic lattice-based rules, here the geometry of the lattice is *not* reflected in the growing structure, unless one zooms in all the way to the scale of the lattice itself.

Growth by branching is readily understood once one realizes that any asperity forming on the growing aggregate tends to capture more random-walking particles than a plane surface. As shown in figure 3.3, on a 2-D Cartesian lattice, there is one and only one way to stick to a plane surface, namely, a step

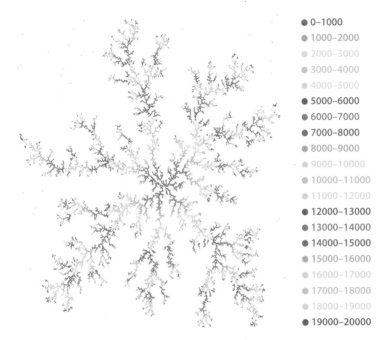

● 0–1000
● 1000–2000
● 2000–3000
● 3000–4000
● 4000–5000
● 5000–6000
● 6000–7000
● 7000–8000
● 8000–9000
● 9000–10000
● 10000–11000
● 11000–12000
● 12000–13000
● 13000–14000
● 14000–15000
● 15000–16000
● 16000–17000
● 17000–18000
● 18000–19000
● 19000–20000

Figure 3.2. Growth of a dendritic structure by DLA. Here 20,000 particles have random walked on a 2-D Cartesian lattice of size 1024×1024, with a single "sticky" seed particle placed at the lattice center at the first iteration. The colors indicate the order in which the free particles have aggregated: red for the first 10^3 particles, orange for the next 10^3, and so on following the color code indicated on the right.

directed perpendicularly toward that surface, as illustrated in figure 3.3A. An asperity, on the other hand, can be reached from many directions, as shown in figure 3.3B, and so will tend to capture more random-walking particles and continue growing. Moreover, once two neighboring dendrites have started to grow, the space in between will be hard to reach, because particles executing a random walk will be more likely to stick to one or the other dendrite, than to reach their branching point. Indeed,

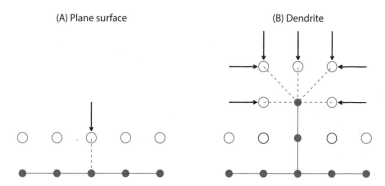

Figure 3.3. Capture of particles (A) on a plane surface, and (B) on the tip of a branch. Linked red solid dots represent aggregated particles, and open red circles lattice nodes on which an incoming particle would be captured. In panel A, each fixed particle controls one such node (red dotted line), itself accessible only from the node vertically above (black arrow). In panel B, in contrast, the aggregated particle at the end of the branch controls five capture sites, which jointly are accessible from seven distinct steps. Note also that the two empty lattice nodes drawn as black open circles in panel B cannot be reached because particles would inevitably stick at a neighboring node one step earlier.

in figure 3.3B the immediate diagonal neighbors of the branching point (open black circles) are simply inaccessible because any particle reaching either of its two neighboring nodes will stick there and proceed no further. The end result is successive growth and branching, rather than homogeneous or statistically uniform filling of the lattice. This effective "exclusion" of nodes neighboring existing branching points is loosely akin to the operation of the one-neighbor-only activation rule used in some of the 2-D CAs investigated in section 2.2.

3.4 A Zoo of Aggregates

So we understand the dendritic shape of aggregates. Yet other factors come into play in determining the type of structure

Figure 3.4. Aggregates in a DLA simulation involving 50,000 particles on a 1024 × 1024 lattice, with 32 randomly distributed particles
assigned "sticky" status prior to the first iteration.

produced. Figures 3.4 and 3.5 show the results of two DLA
simulations, again on a regular 2-D 1024 × 1024 lattice, this
time with 32 sticking particles introduced at random locations on
the first iteration of the simulation. The two simulations differ
only in the number of moving particles placed on the lattice:
5×10^4 for figure 3.4 and 4 times more for figure 3.5.

In figure 3.4, 30 aggregates of varying sizes and shapes can
be counted. This is 2 fewer than the initial number of sticky
particles, a consequence of two "fusion" events between pairs of

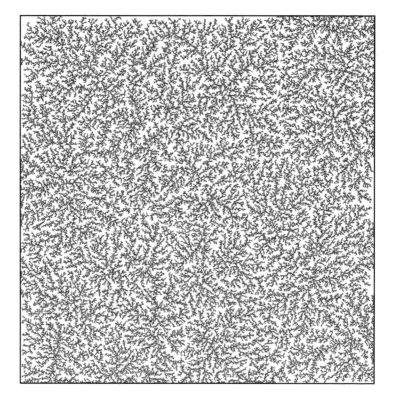

Figure 3.5. The same simulation as in figure 3.4 but for 4 times more particles (200,000). There are 25 individual aggregates here, 7 fewer than the initial 32 sticky particles, because of fusion between some growing aggregates in the course of the simulation. Try pedaling your way out of that maze!

growing aggregates taking place early on in the course of the simulation. Each aggregate shows the same overall branching structure as in figure 3.2, but now their global shape is less "circular." This is because of the finite number of particles available to sustain growth; aggregates growing close to one another will "compete" for the available supply of random-walking particles along the direction linking their geometrical centers. As a result, aggregates will grow preferentially in directions where no other aggregates are located. The close group of 3 aggregates at midheight along the left edge of figure 3.4 offers a nice illustration of this pattern. For the same reason, aggregates growing in (relative) isolation will tend to reach a larger final size. Similar asymmetric growth is observed in many biological organisms, such as sponges or corals, with growth taking place preferentially in directions of greater nutrient concentration.

Although one would be hard pressed to ascertain this visually, there are 25 individual aggregates in figure 3.5. Here, the initial density of random-walking particles is quite high: 200,000 for 1024×1024 lattice nodes, meaning that about one node in five initially contains a particle. Growth then proceeds very quickly, but even in this case the resulting structures retain the dendritic shape characteristic of aggregates generated at lower densities.

The aggregates resulting from the DLA process are not just visually spectacular; they also possess some rather peculiar geometrical properties, most notably *self-similarity* and *scale invariance*. Investigating these properties will first require a detour through fractal geometry, to which we now turn.

3.5 Fractal Geometry

Consider the iterated growth procedure illustrated in figure 3.6. Starting with a seed line segment of unit length ($n = 0$; top), divide this segment into three sections of equal length. Raise an equilateral triangle from the middle segment, as shown on the $n = 1$ curve. Now repeat this process for the four line segments of this $n = 1$ curve, thus leading to the $n = 2$ curve, and so on for $n = 3$, $n = 4$, etc., as shown in figure 3.6 up to $n = 6$. The curve resulting from this process as n keeps increasing is known as the *Koch fractal*.[2]

---

[2]If an equilateral triangle is used as a seed, yet another pretty kind of synthetic snowflake is produced: the *Koch snowflake*. Try it!

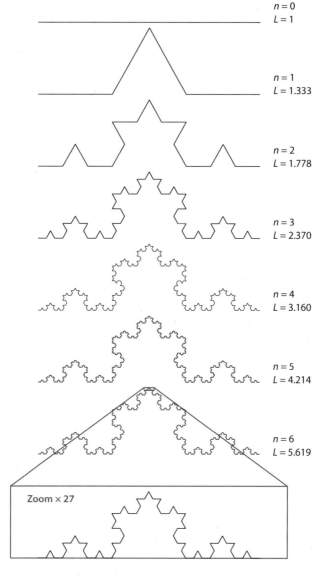

$n = 0$
$L = 1$

$n = 1$
$L = 1.333$

$n = 2$
$L = 1.778$

$n = 3$
$L = 2.370$

$n = 4$
$L = 3.160$

$n = 5$
$L = 4.214$

$n = 6$
$L = 5.619$

Zoom × 27

Figure 3.6. The first six iterations in the "growth" of the Koch fractal. The seed ($n = 0$) is a straight-line segment spanning $[0, 1]$. At each iteration, an equilateral triangle is raised from the central

The Koch fractal is visually pretty, but it also possesses some rather peculiar mathematical properties. The length of each straight-line segment decreases by a factor of 3 at each iteration, as per the rules of the growth process; but the number of these line segments increases by a factor of 4 at each iteration. Consequently, the total length L of the curve increases with the iteration count n as

$$L(n) = 4^n \times \left(\frac{1}{3}\right)^n = \left(\frac{4}{3}\right)^n. \qquad (3.1)$$

Since $4/3 > 1$, the length of the curve will diverge to infinity as n increases; expressed mathematically,

$$\lim_{n \to \infty} L(n) = \infty. \qquad (3.2)$$

Inspection of figure 3.6 may suggest that this divergence is rather slow, and therefore irrelevant in practice—after all, infinity is much farther away than anyone can think. Still, it is an easy exercise to calculate that, starting from a seed line segment of length $L(0) = 5\,\text{cm}$, at $n = 100$ the "unfolded" Koch fractal is already long enough to wrap around the Earth about 4000 times! Now, try to think this through; the infinitely long Koch fractal has well-defined start and end points, namely, the two ends of the original seed line segment. How can an infinitely long line have a beginning and an end? Moreover, this infinitely long line is contained within the definitely finite-sized page of this book. How can an infinitely long line be circumscribed within a geometrical figure like a rectangle, which has a finite perimeter?

Figure 3.6 (*Continued*). third of the line segment, toward the "outside" of the structure. The length L of each curve is indicated on the right. The bottom image is a $27\times$ zoom on the small region delimited by the red rectangle on the $n = 6$ iteration. Note how this zoom is identical to the $n = 3$ iteration.

Such mathematical monsters, as the mathematician Helge von Koch used to call the fractal that now bears his name, cannot be casually dismissed as such because they do occur in the natural world. The fluid dynamicist Lewis Fry Richardson found out the hard way when he tried to measure the length of the British coastline. Working with topographic maps of decreasing scale, he had to come to grips with the fact that the total measured length of the coastline just kept increasing as the map scale decreased, instead of converging to a finite value as he was originally expecting. Yet Britain is most definitely an island, with a clearly finite surface area; a finite surface area bounded by an infinitely long perimeter. Welcome to the bewildering world of fractal geometry!

Loosely speaking, the Koch fractal and the British coastline are "more" than lines, but "less" than surfaces. Geometrically, they are thus objects which should be assigned a dimension between 1 and 2, i.e., a fractional dimension; thus the name "fractal."

Now back to DLA; no matter how complex its shape, the aggregate of figure 3.2 is made up of a finite number of individual particles located in a plane, so each particle can be tagged by two numbers, for example its line and column integer indices on the lattice. On the basis of this parametric definition of dimensionality, it must therefore be declared a 2-D object; so would the CA-generated structures of figures 2.5 and 2.9. This would also be true if the particles were packed in the shape of a square. Yet the aggregate really does not look anything like a solid square, or a pancake, or whatever. The challenge is thus to find a way to quantify this difference.

Consider the two simple geometrical objects illustrated in figure 3.7: a line and a square. Both are constructed by placing a finite number of particles (in red) on a 2-D Cartesian grid similar to that used for the above DLA simulations. It is only on the *macroscopic scale*, much larger than the *microscopic scale* defined by the lattice spacing, that these two objects can be

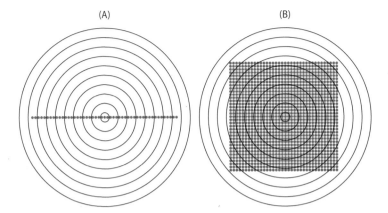

Figure 3.7. The mass–radius method for determining the dimensions of two objects having geometrically simple shapes at the global scale, but made up of individual particles at the microscopic scale: (A) a line, and (B) a solid square. The mass $M(R)$ is defined here as the number of particles contained in a circle of radius R centered on each object. Note how, on each plot, the mass returned for the two outermost circles would be the same, indicating that the global scale of the objects has been reached (see text).

called "line" and "square."[3] Now introduce the same procedure as was used in section 1.2 to evaluate the number density of randomly distributed particles: from the geometrical center of each structure, draw a series of concentric circles of increasing radii R. For each of these circles, count the number of particles it contains, and call this the "mass," denoted hereafter by M. Obviously M increases with R. For a straight line of contiguous particles, as in figure 3.7A, M would grow linearly with R, while for a solid square, as in figure 3.7B, the growth would be quadratic, i.e., $M \propto R^2$. In both cases, this growth can be

[3]To gain an intuitive grasp of this distinction, step back and look at figure 3.7 from an increasing distance and see how far you need to stand to "lose" the granularity of these two geometrical objects.

expressed as a *power law*:

$$M(R) \propto R^D, \quad D \geq 0, \tag{3.3}$$

with $D = 1$ for a line of particles, and $D = 2$ for a solid square. The power-law index D thus provides a measure of the object's dimensionality. Note that equation (3.3) can be expected to hold only for radii significantly larger than the interparticle distance in figure 3.7, and smaller than the global scale of the objects.

Figure 3.8 shows what happens when this mass–radius method is applied to the aggregate of figure 3.2, with the circle's center coinciding with the original sticky particle used to seed the aggregate. The axes being logarithmic, the linear relationship holding in the gray-shaded area indicates that mass still increases as some power of the circle radius.[4] This power law holds well for spatial scales smaller than the size of the aggregate, but significantly larger than the distance between two particles, as set by the lattice spacing. This time the logarithmic slope is $D = 1.665$; even though the aggregate has grown on a 2-D plane, it has a spatial dimension between 1 and 2. In other words, it is "more" than a line but "less" than a surface: again a fractal!

The mass–radius method for determining the fractal dimension is trickier to apply to objects which do not have a well-defined geometrical center. A more robust method is *box counting*, which is particularly appropriate to structures defined on lattices or as pixelated images. Box counting operates as follows. Imagine trying to cover the aggregate of figure 3.2 with a tiling of contiguous squares of size $M \times M$, the measuring unit here being the internodal distance on the lattice (i.e., $M = 8$ means a square

---

[4]Start with the power-law relation $M/M_0 = R^D$; taking the logarithm on both sides yields $\log(M/M_0) \equiv \log M - \log M_0 = \log R^D \equiv D \log R$, so that

$$\log M = D \log R + \log M_0,$$

which is a linear relationship between $\log M$ and $\log R$, with D as the slope and $\log M_0$ as the intercept.

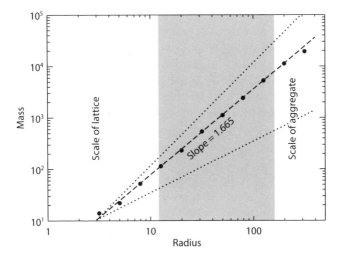

Figure 3.8. Mass–radius relationship for the DLA aggregate of figure 3.2. The logarithmic slope is now 1.665, which, geometrically speaking, places this structure between the "line" and the "square." In other words it is somewhere between a 1-D object ($D = 1$; lower dotted line) and a 2-D object ($D = 2$; upper dotted line). The gray-shaded area indicates the range used to compute the slope (see text).

covering an 8×8 block of nodes). Figure 3.9 illustrates this procedure, for box sizes of $M = 8$, 16, 32, and 64. Now, for each value of M, count the number $B(M)$ of such boxes required to cover the aggregate. Whether a box covers one or many particles making up the aggregate, it always contributes $+1$ to the box count. The smallest meaningful box size is $M = 1$, in which case the count is equal to the number of particles making up the aggregate. The largest meaningful box size is of the order of the linear size of the aggregate; any larger box size would always return a box count $B = 1$, independently of the box size.

Figure 3.10 is an example of a user-defined Python function which performs a box-count calculation on a 2-D array `grid` of size $N \times N$, provided through its argument list. Upon successful

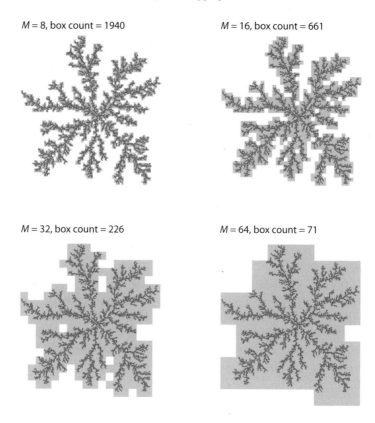

$M = 8$, box count = 1940 $M = 16$, box count = 661

$M = 32$, box count = 226 $M = 64$, box count = 71

Figure 3.9. Four successive doubling steps of the box-counting method, as applied to the aggregate of figure 3.2. Each iteration doubles the linear size M of the gray squares used to cover the structure.

completion the function returns three quantities: the number of scales used for the analysis (the integer n_scales), and two arrays of this size holding the scale size M in nodal units (array scale) and the corresponding box count B (array n_box). This could be called directly at the end of the DLA code presented in figure 3.1 via the following instruction:

```
n_scales,scale,n_box=boxcount(N,grid,2)
```

```
1   # BOX COUNTING FUNCTION FOR FRACTAL INDEX CALCULATION
2   def boxcount(n,grid,occ_val):
3   #  Input is 2D array "grid", of size nxn; value =2 means occupied node
4       n_scales=1                          # Calculate number of scales
5       while (2**n_scales < n) and (n_scales < 100): n_scales+=1
6       scale=np.zeros(n_scales)            # Will hold all box size values
7       n_box=np.zeros(n_scales)            # Will hold the boxcount
8
9       for iscale in range(0,n_scales):    # Loop over allowed scales
10          block_size=2**(iscale+1)        # Block size for this scale
11          n_block=n//block_size           # Number of blocks for this scale
12          n_box[iscale]=0
13          for i in range(0,n_block):      # Loop over first dimension
14              i1=block_size*i             # i-range of this block
15              i2=block_size*(i+1)
16              for j in range(0,n_block):  # Loop over second dimension
17                  j1=block_size*j         # j-range of this block
18                  j2=block_size*(j+1)
19                  if occ_val in grid[i1:i2,j1:j2]: # At least 1 occupied node
20                      n_box[iscale]+=1    # Increment box count
21          # End of lattice loops
22          scale[iscale]=block_size
23          print("scale {0}, boxcount {1}.".format(scale[iscale],n_box[iscale]))
24      # End of scale loop
25      plt.scatter(1./scale,n_box)         # Simple version of Fig 3.11
26      plt.xscale('log')                   # logarithmic axes
27      plt.yscale('log')
28      plt.show()
29      return n_scales,scale,n_box
30  # END FUNCTION BOXCOUNT
```

Figure 3.10. A user-defined Python function performing a box-count calculation on an input array `grid` of size N×N, passed through the function's argument list. Sites having values `occ_val`, passed also as an argument, are deemed occupied for the purpose of a box count. For the DLA simulations produced using the code in figure 3.1, the value 2 should be used. The number of allowed scales (`n_scales`) is first calculated, as the highest power of 2 that yields a value smaller than or equal to the grid size (lines 4–5). The maximum scale number is hardwired at 100; if you fancy doing calculations on lattices of linear size exceeding 2^{100}, then (1) increase this number, and (2) see you in the 22nd century (maybe!). The function is structured around a loop on all allowed scales (starting on line 9), within which two nested loops over the two lattice dimensions build the appropriate blocks (lines 13–18), as the index ranges over the `grid` array. The function returns arrays `scale` and `n_box`, both of length `n_scales`, containing respectively the scales M and corresponding box count $B(M)$. The generalization to rectangular

Now onto the fractal dimension. Figure 3.11 plots the box count as a function of resolution r ($= 1/M$), defined as the inverse of the scale of measurement M (i.e., high resolution $\equiv$ small measuring scale). Logarithmic axes are used once again, and the straight-line fit indicates that the box count is related to the resolution via a power law, here of the form

$$N(r) \propto r^D, \quad D = 1.591. \tag{3.4}$$

This again holds over a range of resolutions, bracketed by the size of the aggregate (r small) and the lattice scale (r large). As before, the logarithmic slope in figure 3.11 directly yields the power-law index D, which is again a measure of the fractal dimension of the aggregate. This version of the fractal dimension is here equal to $D = 1.591$.

Should we be concerned that the fractal dimensions obtained from the mass–radius relation ($D = 1.665$) differs from that extracted from box counting ($D = 1.591$)? Not really. Whatever method is used, here it pertains to a specific aggregate produced by an equally specific realization of the DLA process. To obtain a truly accurate determination of the fractal dimension of aggregates, in general, one would need to generate many such aggregates through statistically independent realizations of the DLA process, combine the box counts, and calculate D. For DLA, the result turns out to be $D = 1.6$, independent of the method used, as it should be. This idea of *ensemble averaging* is discussed in more detail in the next chapter.

Figure 3.10 (*Continued*). arrays, or arrays of dimensions larger than 2, is simple and thus left as an exercise, as the saying goes. The matplotlib plotting instructions on lines 25–28 produce a simplified version of figure 3.11 prior to exiting the function. A linear least-squares fit to this log-log plot allows a determination of the fractal index (see text).

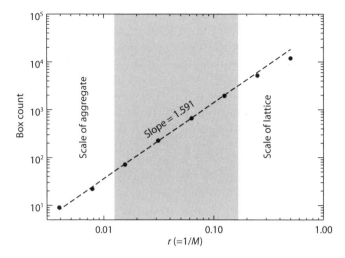

Figure 3.11. Determination of the fractal dimension of the aggregate of figure 3.2 by the box-counting method. As before, the fractal dimension is given by the logarithmic slope of the box count B versus r, as determined over a range of resolutions ($r = 1/M$), bracketed by the size of the structure (small r) and the lattice interval (large r).

3.6 Self-Similarity and Scale Invariance

The defining characteristics of fractal geometry are *self-similarity* and *scale invariance*. Loosely speaking, this means that a fractal structure always "looks the same" upon zooming closer and closer in. We have encountered this already, in section 1.3 with the bifurcation diagram for the logistic map (figure 1.2), with the CA of figure 2.2, as well as with the Koch fractal of figure 3.6. Figure 3.12 illustrates this effect, for our now familiar aggregate of figure 3.2. No matter what the zooming level is, one just sees irregular branches giving rise to more irregular branches, themselves spawning more smaller branches, all the way to the scale of the lattice. Only at that scale can it be clearly perceived that the simulation is carried out on a Cartesian lattice, with sticking to the 8 nearest neighbors, i.e., vertically, horizontally,

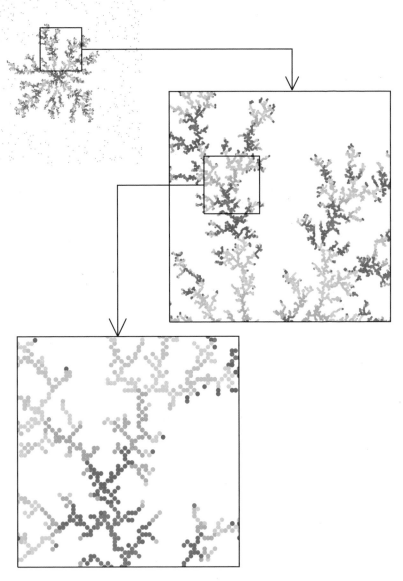

Figure 3.12. Self-similarity in the aggregate of figure 3.2. The two successive zooms each magnify by a factor of 4 in linear size. The color coding indicates the order in which the particles have aggregated, following the color scheme specified in figure 3.2.

and diagonally. Of course, scale invariance also breaks down at the global scale of the aggregate (top left), where a "growing center" is readily identified, and the finite size of the structure becomes apparent.

The break in scale invariance at the smallest and largest spatial scales characterizing the structure is quite typical. It has already showed up in figures 3.8 and 3.11, in the departure of the measurement data points from the power-law relationships. The range in which this relationship holds effectively defines the scale-invariant regime. Indeed, the very existence of a power-law regime in the distribution of some measure of a structure is usually taken as an indicator of scale invariance.[5] But what is responsible for scale invariance? This is a complex (!) question, to which we shall often return in later chapters. In the DLA context, scale invariance reflects *self-similarity* in the growth process: branches grow by spawning more branches, through a sticking process that operates locally and "knows" nothing about the global properties of the growing aggregate.

Nature is replete with scale-invariant structures hard to describe using conventional Euclidean geometry, unlike most technological constructs. A car engine fully taken apart will yield a lot of flat or curved plates, cylinders, disks, rods, pierced hexagons, and so on; now, try to build a snail shell out of regular 2×4 Lego blocks,[6] and while doing so you should have ample time to reflect

---

[5]At a purely mathematical level, a power law is said to be scale invariant for the following reason: Start with a generic power law $f(x) = f_0 x^{-\alpha}$ and introduce a new scale of measurement $x' = ax$ (this could be as simplistic as switching from centimeters to meters as a unit for x, in which case $a = 10^{-2}$). Then we have

$$f(x')/f_0 = (x')^{-\alpha} = (ax)^{-\alpha} = a^{-\alpha} x^{-\alpha}.$$

Defining $f_0' = f_0 a^{-\alpha}$, the power law remains of the form $f(x')/f_0' = x^{-\alpha}$, i.e., neither the power-law form nor the index have been altered by the change of scale.

[6]Readers of a younger generation may try to pick up this challenge on Minecraft instead.

upon the fundamental differences between these two geometrical classes of objects.

3.7 Exercises and Further Computational Explorations

1. Distribute particles on a Cartesian lattice (a) along a line, and (b) filling a square block, as in figure 3.7. Apply the mass–radius method to these two objects, produce plots similar to figure 3.8, and verify that $D = 1$ in the former case, and $D = 2$ in the latter.

2. A simple modification to the DLA code of figure 3.1 can greatly increase its run-time speed. The idea is to inject a single particle per iteration, at some randomly chosen location on a circle circumscribing the growing DLA. You need to implement the following modifications:

    a. Initialization consists in placing a single sticky particle at lattice center $(x, y) = (N/2, N/2)$; set the circle radius to $R = 2$.

    b. At each iteration pick a random angle $\theta \in [0, 2\pi]$ and place a single random-walking particle on the lattice node closest to the position $(x, y) = (N/2 + R\cos\theta, N/2 + R\sin\theta)$.

    c. If a particle sticks, calculate its distance d with respect to the initial, central, sticky particle; if this distance is larger than R, reset $R = d + 1$.

    d. Once the injection circle hits the sides of the lattice, stop injecting particles but keep running the simulation until all remaining random-walking particles have aggregated.

3. Use the Python code of figure 3.1 to explore the effect of varying the initial particle density, the latter defined as the total number of random-walking particles divided by the total number of available lattice nodes.

At what density can you finally produce an amorphous solid object? Using the box-counting method on your sequence of aggregates, determine whether or not their fractal dimension is influenced by the initial particle density.

4. Grow some aggregates starting from a row of sticking particles located along one edge of the lattice. Do so for a uniform random initial distribution of moving particles and for a Gaussian initial distribution centered on the middle of the lattice (if needed, see appendix C on how to produce Gaussian distributions of random deviates). Experiment with different values for the density of random-walking particles, or for the location of the initially sticky particles.

5. Modify the Python code of figure 3.1 (or better, the alternative version you built in exercise 2) so that moving particles stick only if they have a sticking particle at one of their four closest neighbors, top/down/right/left, i.e., excluding diagonal neighbors. Using a single sticking particle as a seed, as in figure 3.2, reflect upon the impact this change has on the overall appearance of the resulting aggregate, and contrast this with the impact of a similar change on the deterministic growth rules used to produce the structures in figure 2.9.

6. And now for the Grand Challenge: set up and carry out a DLA simulation on a 6-neighbor triangular lattice, starting from a single sticky particle at the lattice center. This really only requires altering the template arrays dx and dy in the code listed in figure 3.1 (or the faster version designed in exercise 2; see also figure 2.5 for inspiration). Determine the fractal dimension of the resulting aggregate (the mass–radius method will be fine here). Is the fractal dimension dependent on the

assumed lattice topology? Think carefully about the best way to apply the mass–radius and/or box-counting methods on such a lattice; you may start by taking yet another look at figure 2.5.

3.8 Further Reading

The DLA model introduced in this chapter essentially follows

Witten, T.A. Jr, and Sanders, L.M., "Diffusion-limited aggregation, a kinetic critical phenomenon," *Phys. Rev. Lett.*, **47**, 1400–1403 (1981).

The Wikipedia page on DLA is rather minimal, but does include a nice photograph of a copper sulfate aggregate grown in the laboratory through DLA (March 2015):

http://en.wikipedia.org/wiki/Diffusion-limited_aggregation.

A multitude of good books are available on fractal geometry. The following is among the most influential early discussions of the subject, and is still well worth reading:

Mandelbrot, B., *The Fractal Geometry of Nature*, Freeman (1982).

At the textbook level, try

Falconer, K., *Fractal Geometry*, John Wiley & Sons (2003).

I also found the following web page quite informative (viewed June 2016):

http://users.math.yale.edu/public_html/People/frame/Fractals/.

I gained much inspiration and insight on naturally occurring fractal geometry from the following two books, which I thus take the liberty to cite even though they may not be the optimal references on the topic:

Prusinkiewicz, P., and Lindenmayer, A., *The Algorithmic Beauty of Plants*, Springer (1990).

Flake, G.W., *The Computational Beauty of Nature*, MIT Press (1998).

4

PERCOLATION

We saw in the preceding two chapters that rule-based growth, whether in CAs or through DLA, can lead to the buildup of complex structures, sometimes exhibiting fractal geometry. We now examine another lattice-based system where similarly complex structures can arise from pure randomness. *Percolation* usually refers to the passage of liquid through a porous or granular medium. In its more abstract form, as developed in this chapter, it has become an exemplar of *criticality*, a concept in statistical physics related to *phase transitions*. An iconic example of the latter is liquid water boiling into water vapor, or freezing into ice.

Superficially, the simple lattice-based model introduced in this chapter bears no relation whatsoever to boiling water or to the flow of fluids through porous media. Yet it does capture the essence of the critical behavior characterizing these systems; such is the power of physical and mathematical abstraction.

4.1 Percolation in One Dimension

Consider a 1-D lattice of length N, i.e., a chain of N nodes each connected to its immediate right and left neighbors, with the exception of the two nodes at the ends of the lattice, which have only one neighbor. Figure 4.1 shows an $N = 64$ example. Each node has a probability p of being occupied (with, of course,

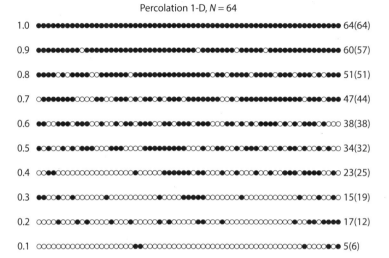

Figure 4.1. Percolation lattices in one spatial dimension. Each line represents an $N = 64$ lattice, with occupied nodes in black and empty nodes left as open circles. The occupation probability p increases from bottom to top in steps of 0.1, as indicated on the left. The number of occupied nodes is listed on the right, followed by the value $p \times N$ expected statistically, within parentheses.

$0 \leq p \leq 1$). This *occupation probability* is the same for all nodes, and is independent of neighboring nodes being empty or occupied; in other words, each node is *statistically independent* of all others on the lattice. For a very large lattice ($N \to \infty$), the expected number of occupied nodes tends toward pN, but at any finite N, deviations from this expected value are anticipated, and may be substantial for small N. This is indeed the case in figure 4.1, where the $p = 0.3$ lattice here contains *fewer* occupied nodes than at $p = 0.2$.

If p is small, only a few nodes on the lattice will be occupied, and most will have empty nearest-neighbor nodes. But as p is increased, the likelihood of having neighboring nodes occupied also increases. Define a *cluster* as a set of contiguous occupied nodes, delineated by one empty node at each end. With p the probability of a node being occupied, then $1 - p$ is the probability of a node being empty. The probability of having at least one cluster of length s is thus

$$(1 - p) \times \underbrace{p \times p \times p \times \cdots}_{s \text{ times}} \times (1 - p) = p^s (1 - p)^2. \quad (4.1)$$

This expression tends toward zero for very large clusters ($s \rightarrow \infty$), even in the limit $p \rightarrow 1$. This reflects the fact that one empty node somewhere is enough to "break" a cluster that is otherwise of length $s \rightarrow \infty$. Nonetheless, at some finite N, the probability of having a cluster of size s increases with p, as one would have expected.

Let s_k measure the size, i.e., the number of occupied nodes, for the kth cluster on the lattice, and denote by S the size of the largest such cluster:[1]

$$S = \max(s_k), \quad k = 0, 1, 2, 3, \ldots. \quad (4.2)$$

Consider now what happens as p is gradually increased. As long as relatively few nodes are occupied, one may expect that existing clusters will grow by "tacking" on a new occupied node at one of their extremities. The probability of this happening increases linearly with p, so one would expect $S \propto p$ for p small. This expectation is borne out in figure 4.1: S grows from 2 to 4 to 5 to 6 as p increases from 0.1 to 0.2 to 0.3 to 0.4. However, once a substantial fraction of lattice nodes are occupied, many clusters of significant size exist on the lattice, and a new growth process emerges: fusion of two preexisting clusters separated by one empty node, once that node becomes occupied. As p continues to increase and the lattice fills up, fusion of ever larger clusters

---

[1] To be consistent with Python's array indexing convention (see appendix A), the K clusters on the lattice are numbered from 0 to $K - 1$. Sorry!

becomes increasingly frequent, and leads to very rapid growth of S. It can be shown that in the limit $N \to \infty$, the size of the largest cluster grows according to

$$\lim_{N \to \infty} S = \frac{1 + p}{1 - p}. \qquad (4.3)$$

This indicates that the size of the largest cluster tends to infinity in the limit $p \to 1$. In other words, the largest clusters reach sizes comparable to that of the whole system. The numerical value of p at which this happens is called the *percolation threshold*, hereafter denoted p_c. For 1-D lattices, $p_c = 1$ for the very simple reason that only one empty lattice node is enough to "break" the infinite cluster. This conclusion would have been easy to anticipate without all this probabilistic mumbo jumbo, but it was important to go through it nonetheless, because things become a lot trickier—and complex!—for lattices in more than one spatial dimension.

4.2 Percolation in Two Dimensions

Let's move to 2-D lattices, and see how much of what we learned in one dimension carries over. In what follows we restrict ourselves to regular Cartesian lattices with top/down/right/left nearest-neighbor connectivity. Each node on the lattice is identified by a pair of integers (i, j) flagging its "vertical" and "horizontal" locations, respectively (if needed, see appendix D for more on lattice definition and notation). Except for nodes located at the lattice boundaries, the 4 nearest neighbors of node (i, j) are

$$\underbrace{(i - 1, j)}_{\text{top}}, \quad \underbrace{(i + 1, j)}_{\text{down}}, \quad \underbrace{(i, j + 1)}_{\text{right}}, \quad \underbrace{(i, j - 1)}_{\text{left}}. \qquad (4.4)$$

Here is a small source code in the Python programming language that defines such a 2-D lattice of size 128×128, and fills it with occupation probability $p = 0.59$.

Figure 4.2. Two-dimensional regular Cartesian lattices of size $N \times N = 64 \times 64$, with occupation probabilities $p = 0.25$, 0.5, and 0.75. Occupied nodes are filled in black, and empty nodes are white.

```
1   # CREATES AND FILLS A 2-D CARTESIAN PERCOLATION LATTICE
2   import numpy as np
3   import matplotlib.pyplot as plt
4   #-------------------------------------------------------------------
5   N=128                           # lattice size
6   p=0.59                          # occupation probability
7   np.random.seed(1234)            # seed for random number generator
8   #-------------------------------------------------------------------
9   lattice=np.zeros([N,N],dtype='int')  # 2-D lattice initialized to zero
10
11  for i in range(0,N):            # lattice loops
12      for j in range(0,N):
13          if np.random.uniform() < p:   # occupy this node
14              lattice[i,j]=1
15  plt.imshow(lattice,interpolation="nearest")
16  plt.show()                      # display lattice
17  # END
```

Note that the value 1 is used to identify an occupied node, empty nodes being set to 0. Clusters are now defined as groups of contiguous occupied nodes separated from other clusters or single occupied nodes by empty nodes. The percolation threshold is now defined as the value of p at which the largest cluster spans the whole lattice, in the sense that it "connects" one lattice boundary to its counterpart on the facing boundary.

Figure 4.2 shows three examples of 2-D regular Cartesian lattices of size $N \times N = 64 \times 64$, with occupation probabilities $p = 0.25$, 0.50, and 0.75. At $p = 0.25$, the lattice contains a large number of small clusters or isolated occupied nodes. Their spatial distribution is random but statistically uniform. It is quite clear here that no single cluster spans the whole lattice, so we are obviously below the percolation threshold. At $p = 0.75$, the lattice looks like a porous object, a bit like a sponge, containing many small holes again distributed randomly but in a statistically uniform manner. Here, one single, very large cluster fills the lattice and contains the majority of occupied nodes. This indicates that we are *beyond* the percolation threshold. The $p = 0.5$ case is more ambiguous, at least visually. Are we seeing a dense clump of large clusters, or a highly fragmented solid structure? That lattice would have to be studied carefully to verify whether or not one cluster extends from one end of the lattice to another. But figure 4.2 already allows us to draw one interesting conclusion: unlike in the 1-D case, here the percolation threshold $p_c < 1$. This is because in two spatial dimensions, an empty node can be bypassed.

4.3 Cluster Sizes

If building a 2-D percolation lattice can be done in a few lines of Python code, identifying and sizing clusters is a much more complex endeavor. There are many algorithms available to do this, and the bibliography at the end of this chapter includes a few good references for those wishing to delve into the state of the art. The algorithm introduced in what follows is far from the most efficient, but it is relatively easy to code and conceptually simple to understand.

Imagine tagging an occupied node with a specific color, say green; starting from this newly colored node, color green all occupied nodes that are nearest neighbors, and then their nearest neighbors, and so on until no uncolored nearest neighbors are found. Then move to the next as-yet-uncolored, occupied node, and repeat this process with a new color tag. Continue in this manner until no uncolored, occupied node is left on the lattice, and each cluster will end up tagged with a unique color.

The Python code in figure 4.3 is a direct implementation of this simple algorithm. This user-defined function could be called directly, at the end of the small code presented at the beginning of this section, and the clusters plotted, through the following instruction:

```
n_cluster,size_cluster,tag_cluster,map_cluster=findcluster(N,lattice)
```

Algorithmically, this function operates along the lines described above:

1. The first step is to copy the N×N lattice into a working array `map_cluster` of size (N+2) × (N+2), thus leaving a padding of unoccupied ghost nodes (value = 0) along its perimeter. This is carried out on line 9, through the implicit looping allowed by the "i1:i2" array index syntax in Python, which means "access elements starting at index value i1 up to position i2 (meaning, index i2-1(!); see appendix A.2 for more on this if needed). This will allow the nearest-neighbor check to be carried out for all nodes using the same relative template, much as in the DLA code of figure 3.1; otherwise, boundary nodes would need to be treated differently, increasing coding complexity. See appendix D.1 for more on the use of ghost nodes.

2. The algorithm is built on two nested outer loops, each running over one dimension of the lattice (lines 12–13), scanning it line by line.

3. At each node scanned within the outer loops, a test verifies whether the node is occupied (value 1) and not

```
1   # FUNCTION FINDCLUSTER: TAGS AND PLOTS PERCOLATION CLUSTERS ON A 2D LATTICE
2   #    lattice is supposed of size NxN with nodal value 1 indicating an
3   #    occupied node and a value 0 for an empty node
4   def findcluster(N,lattice):
5       dx,dy=np.array([-1,0,1,0]),np.array([0,-1,0,1])   # Template arrays
6       size_cluster=np.zeros(N*N/2,dtype='int')          # Cluster size array
7       tag_cluster =np.zeros(N*N/2,dtype='int')          # Cluster tag array
8       map_cluster =np.zeros([N+2,N+2],dtype='int')      # Cluster map array
9       map_cluster[1:N+1,1:N+1]=lattice[:,:]             # Pad lattice with zeros
10      n_cluster,iic=0,100                               # Counter, first tag
11
12      for j in range(1,N+1):                            # Outer lattice scan
13          for k in range(1,N+1):
14              size,add_to_size=0,0                      # Initialize counters
15
16              if map_cluster[j,k] == 1:                 # Initiate new tagging
17                  map_cluster[j,k]=iic                  # New cluster tag
18                  size+=1                               # First node of cluster
19                  add_to_size+=1
20
21                  while( add_to_size > 0)               # Tagging in progress
22                      add_to_size=0
23                      j1,j2=j,min(N,j+size)             # Range of inner scan
24                      k1,k2=max(1,k-size),min(N,k+size)
25                      for jj in range(j1,j2+1):         # Inner lattice scan
26                          for kk in range(k1,k2+1):
27                              if map_cluster[jj,kk] == 1:   # Untagged occupied node
28                                  if iic in map_cluster[jj+dx[:],kk+dy[:]]:
29                                      map_cluster[jj,kk]=iic  # assign tag to node
30                                      size+=1
31                                      add_to_size+=1
32                  # end of inner lattice scan
33                  size_cluster[n_cluster]=size          # Size of this cluster
34                  tag_cluster[n_cluster] =iic           # Tag for this cluster
35                  print("cluster tag {}, size {}.".format(iic,size))
36                  iic=np.random.random_integers(10,250) # Set up next tag
37                  n_cluster+=1                          # Increment counter
38              # end of this tagging
39
40      # end of outer lattice scan; display clusters
41      plt.imshow(map_cluster,interpolation="nearest") # Display clusters
42      plt.show()
43      return n_cluster,size_cluster,tag_cluster,map_cluster
44   # END FUNCTION FINDCLUSTER
```

Figure 4.3. A user-defined Python function for identifying and tagging individual clusters in a 2-D percolation lattice with 4-neighbor connectivity. The matplotlib instructions on lines 41–42 display the clusters essentially as in figure 4.5.

yet assigned to a cluster (line 16). If so, a unique numerical tag (variable `iic`) is assigned to it (line 17).

4. If and only if a new tag has been generated, a new "inner" lattice scan is initiated (line 21–32). Each occupied node having a nearest neighbor with identifier `iic` is tagged with that same identifier (lines 27–29). A `while` loop construct (starting at line 21) ensures that the inner lattice scan is repeated until no untagged, occupied node is found with an `iic`-tagged neighbor in the course of a complete scan.

5. By the design of the algorithm, the cluster being tagged can enlarge only by one nodal distance horizontally and/or vertically from its starting node at each iteration of the `while` loop. Consequently, the inner lattice scan spans an increasing range of nodes with each iteration (lines 23–24), with the use of `min/max` to avoid out-of-bounds array indexing on the array `map_cluster`. Note also that the order in which the lattice is scanned implies that all nodes with index $jj < j$ have already been tagged, so that the range of the inner loop on line 25 begins at `j1=j`.

6. The nearest-neighbor check uses the two 4-neighbor template arrays `dx` and `dy`, verifying whether any nearest neighbor has already been tagged with the value `iic` (line 28). Note here the use of the Python-specific construct `if iic in...`, which means "if value `iic` is found in the set of array elements following"; if needed, see appendix A.5 for an equivalent set of Python instructions using only simple `for` and `if` instructions.

7. At the end of the inner lattice scan, the number of nodes tagged with value `iic` is stored in the array `size_cluster` (line 33), and the outer lattice scan resumes from where it had been interrupted, until a

new untagged, occupied node is located, in which case step 3 begins anew, or until the outer scan reaches the end of the lattice.

8. At the end of the outer lattice scan, the integer variable n_cluster contains the number of clusters identified, the array size_cluster contains the size (measured in number of nodes) of each of these clusters, in the order of their tagging, and the array tag_cluster contains the corresponding numerical value of the tags. Nodal values in the lattice array map_cluster now contain, at occupied nodes, the tag value iic associated with each cluster, instead of the value 1 originally indicating an occupied node (as per line 29). These are the quantities returned by the function (line 43).

9. The size_cluster and tag_cluster arrays are assigned a length of $N^2/2$ (lines 6–7), which is equal to the largest number of clusters than can be fit on an $N \times N$ lattice, namely, clusters all of size 1 distributed as a checkerboard pattern.

Figure 4.4 illustrates the operation of this cluster-tagging algorithm, here for a small 16×16 lattice at occupation probability $p = 0.58$. In the top-left frame, five clusters have already been tagged, as indicated by distinct colors, and the 12 frames cover successive tagging steps (iterations of the while loop) within the outer lattice loop, starting from a untagged, occupied node at the upper left (in green).

This algorithm is (relatively) easy to code but inefficient in a number of ways, notably the fact that the outer and inner sets of loops spend a lot of time revisiting nodes that are unoccupied or have already been tagged. A more efficient approach, relatively straightforward to code in Python, would be to first build a *list* of occupied nodes, and replace the two sets of loops at lines 12–13 and 25–26 by a single loop over elements of that list. Elements of

Figure 4.4. The cluster-tagging algorithm of figure 4.3 in action, here on a 16×16 lattice at $p = 0.58$. The nodal position $(0, 0)$ is at the top left in each frame. The outer and inner lattice scans proceed, line by line, from top to bottom (loop indices j and jj) in the code of figure 4.3, and from left to right in each line (loop indices k and kk). Colored squares correspond to occupied nodes already tagged to a cluster, while gray squares indicate as-yet-untagged occupied nodes, and white squares indicate empty nodes. In the top-left frame, a new tagging inner loop has just started at the left extremity of the third line of the lattice (nodal position $(j, k) = (2, 0)$), and the bottom-neighbor node has just been tagged (both in green). The next 11 frames show successive steps of the tagging process, each corresponding to an iteration of the while loop at line 21 in figure 4.3, the sequence being obvious. At the end of the tagging process (bottom right), a cluster of 44 nodes has been tagged "green." The outer lattice scan would now resume in the third column, with the node $(j, k) = (2, 1)$. The next tag would be initiated at node $(2, 6)$, for a cluster of size 2.

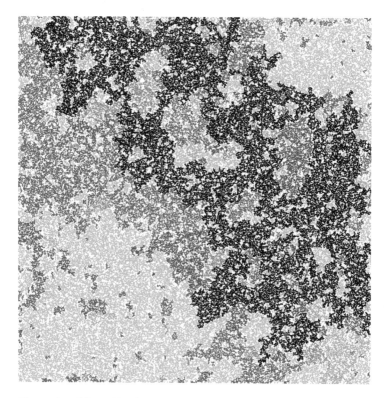

Figure 4.5. The 661 largest clusters on a 512×512 lattice at $p_c = 0.59$. Empty sites are left white, while gray indicates occupied nodes that are not part of one of the 661 largest clusters. The largest cluster, plotted in black, collects $S = 53,537$ of the 154,867 occupied nodes, and spans the whole lattice. Notice how holes in the larger clusters contain smaller clusters, themselves with holes containing even smaller clusters, and so on down to single occupied nodes. The matplotlib instructions at the end of the cluster-tagging function in figure 4.3 generate essentially this type of display.

the array grid are tagged as before, but nodes are then removed from the list as they are tagged.

Figure 4.5 shows the end result of the tagging algorithm of figure 4.3, here for a 512×512 lattice at $p = 0.59$. Upon completion of the tagging algorithm, locating and tracing the

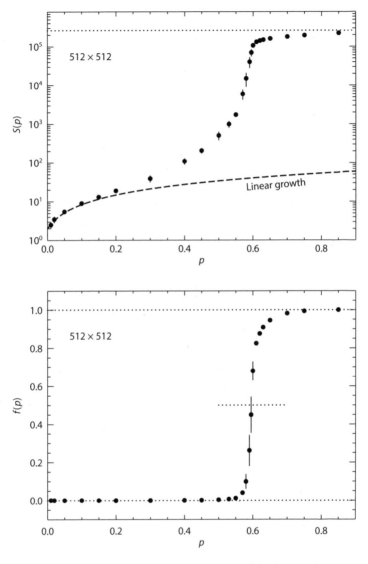

Figure 4.6. The top panel shows the growth of the largest cluster on a 512×512 lattice, as a function of the occupation probability p. Note the logarithmic vertical axis. Each solid dot is an average over 10 statistically independent realizations of the lattice at the same value of p, with the vertical line segments indicating the standard

largest cluster simply requires searching the array `size_cluster` for its largest element, retrieving the associated tag number from the array `tag_cluster`, and finally extracting the correspondingly numbered nodes from the cluster map array `map_cluster`. These jointly form the largest cluster, colored in black in figure 4.5. The top panel in figure 4.6 shows how the size S of that largest cluster increases with the occupation probability p, still for an $N \times N = 512 \times 512$ lattice. What is plotted is actually the mean largest cluster size $\langle S \rangle$, averaged over $M = 10$ realizations of the lattice at each value of p (solid dots), with the vertical line segments indicating the standard deviation σ_S about the mean:

$$\langle S \rangle = \frac{1}{M} \sum_{m=1}^{M} S_m, \quad \sigma_S = \left(\frac{1}{M} \sum_{m=1}^{M} (S_m - \langle S \rangle)^2 \right)^{1/2} \quad (4.5)$$

As when computing the fractal dimension of DLAs in the preceding chapter, such *ensemble averaging* is carried out to ensure that the plotted variation is representative, and not distorted by the idiosyncrasies of a specific lattice configuration, each percolation lattice being as unique as the seed provided to the random number generator upon initialization (if needed, see appendix C.2 on this seed business). As expected, the size S of the largest cluster grows with the occupation probability.[2] For

---

[2]Note that throughout this chapter, the term "growth" is used even though it does not arise from the action of a dynamical process, such as in chapters 2 and 3.

Figure 4.6 (*Continued*). deviation σ_S about the the ensemble mean $\langle S \rangle$ (see equations (4.5)). The dashed curve corresponds to linear growth, and the dotted line indicates the largest cluster size possible on the lattice, here $512 \times 512 = 2.62 \times 10^5$. The bottom panel plots the same numerical results, but for S normalized by the total number of occupied nodes (see equation (4.6)), and with the vertical axis now linear rather than logarithmic.

$p \lesssim 0.2$, linear growth (dashed curve) fits the numerical data tolerably well, but in the range $0.2 \lesssim p \lesssim 0.6$ growth has already become super-exponential (i.e., upward curvature in this log-lin plot). This reflects successive pairwise fusion of existing clusters, through the occupation of single nodes that had remained empty at lower p values. Sometimes, occupying just one more node is all it takes. The rapid saturation at $p \gtrsim 0.6$ is set by the size of the lattice, which limits here the largest cluster to a maximal size of 512×512 (dotted line).

The bottom panel in figure 4.6 shows the same results as in the top panel, except that now the size of the largest cluster has been normalized by the expected number of occupied nodes on the lattice at each value of p, i.e.,

$$F(p) = \frac{S(p)}{p N^2}. \tag{4.6}$$

This measures the *fraction* of occupied nodes belonging to the largest cluster, and it highlights something interesting: as long as $p \lesssim 0.5$, $F(p)$ remains close to zero, even though the absolute size of the largest cluster is growing significantly (cf. top panel, and its logarithmic vertical axis!). In other words, the largest cluster becomes bigger, but does not particularly stand out as compared to other clusters on the lattice. At the end of the range, $p \gtrsim 0.65$, the largest cluster includes almost all occupied nodes, which we expected already. But what is striking is the sharpness of the transition between these two regimes. Around $p = 0.55$, $F(p)$ grows very rapidly, already approaching saturation close to unity at $p \simeq 0.65$. Indeed, around $p = 0.6$, the growth of S appears to diverge, in the (calculus) sense that $dF/dp \to \infty$. The exact value of p at which this takes place defines the percolation threshold for this 2-D lattice. At this threshold, the largest cluster contains on average half of the occupied nodes:

$$S(p_c) = \tfrac{1}{2} p_c N^2. \tag{4.7}$$

For a 4-nearest-neighbor 2-D Cartesian lattice, the percolation threshold turns out to be at $p_c = 0.592746$. Unlike in the 1-D case, there exists no equivalent to equation (4.3), and the percolation threshold must be evaluated numerically.

Enough (for now) with the largest cluster, and let's turn to the population of all clusters on the lattice. This information is contained in the array `size_cluster` returned by the cluster-tagging code listed in figure 4.3. We now want to get a measure for the range of cluster sizes found in the lattice. To this end, the most useful mathematical object is a *probability density function* (hereafter, often abbreviated as PDF) of cluster sizes.

Mathematically, the PDF $f(s)$ is defined such that $f(s)\Delta s$ gives the probability of finding on the lattice a cluster of size between s and $s + \Delta s$. Figure 4.7 plots such PDFs of cluster sizes on an $N \times N = 512 \times 512$ lattice with $p = 0.3$ (red), 0.59 (green), and 0.7 (blue). In essence, these discrete PDFs thus measure, in each simulation, the frequencies of clusters having a size that falls within each of the histogram bins.[3] Like most PDFs to be encountered later in this book, the PDFs in figure 4.7 are plotted in so-called histogram mode, to emphasize their fundamentally discrete nature: a count of clusters in a given size range Δs is an integer number and characterizes a finite-sized range.

In the first case, $p = 0.3$, the PDF drops rapidly as s increases, reflecting the fact that the lattice is populated by small clusters (as in figure 4.2, left panel). At $p = 0.7$, one gigantic cluster contains nearly all occupied nodes (as in figure 4.2, right panel). This single supercluster accounts for the single blue histogram column at $s \simeq 2 \times 10^5$. The remaining clusters are small ones dispersed in

---

[3]Readers unfamiliar with this concept should really read appendix B before proceeding any further. Note also that the PDFs plotted in figure 4.7 are constructed using logarithmically constant bin sizes Δs, as described in appendix B.5.

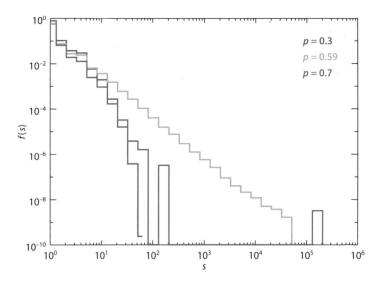

Figure 4.7. Probability density functions of cluster sizes on a 512×512 lattice for $p = 0.3$ (red), 0.59 (green), and 0.7 (blue). Note the similarity between the PDFs at $p = 0.3$ and $p = 0.7$, the crucial difference being the presence of the lone blue histogram bin at $s \simeq 2 \times 10^5$, corresponding to the single largest cluster covering most of the lattice at $p = 0.7$.

the cavities of the supercluster, and their PDF closely resembles that of clusters at $p = 0.3$. The case $p = 0.59$ is very close to the percolation threshold, and stands out in that its PDF takes the form of a power law spanning essentially the whole range of cluster sizes accessible on the lattice:

$$f(s) = f_0 s^{-\alpha}, \quad \alpha > 0, \tag{4.8}$$

here with $\alpha \simeq 1.85$. What is truly remarkable is that the numerical value of this exponent is independent of lattice size, as shown in figure 4.8. In going from small lattices of a few

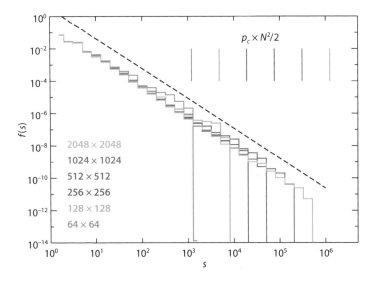

Figure 4.8. Cluster-size PDFs near the percolation threshold, for 10-member ensembles of statistically independent realizations of lattices ranging in size from $N \times N = 64 \times 64$ to 2048×2048, as color coded. The vertical tick marks in the upper right indicate the expected size ($S = p_c N^2 / 2$) for the largest cluster on each lattice, and the dashed line is drawn with a logarithmic slope of -1.85, which on such log-log plots gives the exponent α in equation (4.8) directly.

thousand nodes, to lattices in excess of a million nodes, the PDF retains the same power-law shape and logarithmic slope; all that changes is the extension of the distribution to ever larger sizes, the cutoff always occurring very near the expected size $p_c N^2 / 2$ for the largest cluster at the percolation threshold. The power-law index $\alpha = 1.85$ is said to be *universal* for this class of 2-D Cartesian lattices with 4-nearest-neighbor connectivity. These PDFs are again ensemble averages of 10 realizations of the percolation lattice at each value of occupation probability.

Each PDF is constructed from combining cluster counts for all 10 realizations, and then the power-law index $\alpha = 1.85$ is calculated for this joint PDF.[4]

4.4 Fractal Clusters

A robust power-law PDF is indicative of scale invariance in the structure being measured. We have already encountered scale invariance in our discussion of fractal geometry in the preceding chapter. Could clusters on a lattice at the percolation threshold be fractal objects? Let us look into that.

Figure 4.9 displays the largest cluster found on a 512×512 lattice, for occupation probabilities ranging from $p = 0.57$ to 0.6, as labeled. The largest cluster is still significantly smaller than the lattice at $p = 0.57$, already spans it at $p = 0.59$, and fills it in a spongelike manner at $p = 0.6$. The shapes of these clusters are noteworthy. Close to the percolation threshold, the clusters are very filamentary and contain many large cavities, which contain smaller clusters also with cavities, also containing smaller clusters, and so on down to the scale of the lattice interval (see figure 4.5), in the same classical scale-invariant manner as in the DLA in figure 3.2. Clusters are indeed fractal objects, with a dimension somewhere between 1 and 2. Their fractal index varies with the occupation probability p, the numerical value being smallest at the percolation threshold. Because of their highly irregular shape, the fractal dimension of clusters is best computed using the box-counting method introduced in section 3.5.

4.5 Is It Really a Power Law?

Power-law PDFs pop up everywhere in measurements of "event sizes" in naturally occurring phenomena: for example, avalanches, forest fires, earthquakes, solar flares, to name but a few that

---

[4]Note that this is *not* the same as averaging the 10 power-law indices associated with the individual PDFs for each member of the ensemble, since a power law is a nonlinear function of its independent variable, which here is size occurrence frequency $f(s)$.

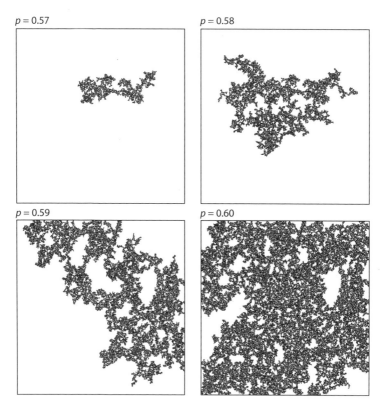

Figure 4.9. The largest clusters on a 512×512 lattice, for occupation probabilities $p = 0.57$, 0.58, 0.59, and 0.6. Over this very restricted range, a small increase in p leads to a pronounced increase in the size of the largest cluster.

will be encountered in subsequent chapters. The implied scale invariance holds important clues to the underlying dynamical processes driving these events, and consequently a reliable empirical determination of power-law form (4.8) and associated index α is important.

The power-law index $\alpha = 1.85$ characterizing the PDFs in figure 4.8 was obtained by a linear least-squares fit on the 2048×2048 PDF in the range $10 \leq s \leq 10^5$, with the fit carried out in the log-log plane and equal weight assigned to each histogram bin. This very simple method has its limitations and must be used with proper caution. If the PDF extends over many orders of magnitude in the measured variable (here over five orders of magnitude in s) and is built from a great number of measured events (here over 10^6 for the 10-member ensemble), the inferred power-law index typically turns out to be fairly accurate; this is often no longer the case for steeper power-law PDFs, spanning only a few orders of magnitude and/or built from a smaller sample of measured events. Some robust statistical approaches have been designed, that allow reliable determination of power-law indices even under these circumstances. See appendix B.6 for more on these matters.

The fractal structure of percolation clusters will certainly not extend to the very smallest cluster sizes possible: a cluster of size 2 is definitely a line; so are one-third of clusters of size 3, the other two-thirds having the shape of 90-degree wedges; and so on. Scale invariance will surely break down before reaching the smallest cluster size, like it did when zooming in on the DLA aggregates (see figure 3.12). Likewise, the finite size of the percolation lattice will inevitably distort the shape of the largest clusters. This effect is clearly visible in figure 4.9 where, close to the percolation threshold ($p = 0.59$; bottom left), parts of the largest cluster are clearly deformed due to the presence of the lattice boundaries. Scale-invariant power-law behavior is thus expected to break down at the high end of the cluster-size distribution as well. This is why the fit in figure 4.8 is carried out using data in the range $10 \leq s \leq 10^5$ only.

4.6 Criticality

Let's summarize what we have learned so far about 2-D percolation. At and only at the percolation threshold p_c, the following holds:

1. The sizes of clusters are distributed as a power law.
2. The linear dimension of the largest cluster is $\simeq N$.

3. The largest cluster collects a fraction $F = 0.5$ of all occupied nodes.

4. The growth rate of $\langle S \rangle$ diverges ($d\langle S \rangle / dp \to \infty$) in the limit $p \to p_c$.

5. The root-mean-squared (rms) deviation of the size of the largest cluster, relative to the mean value, is largest.

6. The fractal dimension of the largest cluster reaches its smallest numerical value.

These behaviors characterize what is known in statistical physics as a *critical system*. The operational defining characteristic of a critical system is its global extreme sensitivity to a small perturbation in the system. Phase transition in water is the typical example, whereby water at 100°C transits from liquid to vapor; there is no such thing as a pot of half-boiling water; either the whole pot is boiling, or it is not, and under so-called standard atmosphere conditions, the transition point is at exactly 100°C. A tiny fraction of a degree below 100, and the water is liquid; a tiny fraction of a degree above, and the water is already vapor. But at exactly 100°C, adding a tiny increment of heat will trigger the phase transition.[5]

The link with percolation is with the behavior of the largest cluster as a function of the occupation probability. When $p < p_c$, adding an occupied node will perhaps enlarge a cluster; when $p > p_c$, there is already a large cluster spanning the lattice, and adding one more occupied node to it will not change much. But at exactly $p = p_c$, adding a single node *may* connect two existing large clusters to generate a cluster spanning the whole lattice. If we think of the latter as a porous medium (occupied node = material, empty node = hole), the system goes suddenly from permeable to impermeable. If the lattice is viewed

---

[5]In the language of statistical physics, one would say that the correlation length of a perturbation becomes comparable to the size of the system.

as some composite material made of electrically conducting grains (occupied nodes) embedded in a nonconducting matrix (empty nodes), then at the percolation threshold the system goes suddenly from nonconducting to electrically conducting. Other well-studied examples include magnetization at the Curie point, polymerization of colloidal liquids, and superfluidity in liquid helium, to name but a few.

In all these systems, critical behavior materializes when a *control parameter* is very finely tuned to a specific value—$p_c = 0.59274$ for percolation on a 4-neighbor 2-D Cartesian lattice; a temperature of 100°C for boiling water, etc.—by a mechanism external to the system. The need for such finely tuned external control may suggest that criticality is unlikely to develop spontaneously in natural systems, which are typically not subjected to finely tuned external control.

It turns out that many natural systems *can* reach a critical state autonomously, through the action of their own dynamics, and, as a matter of fact, the following chapter introduces one.

4.7 Exercises and Further Computational Explorations

1. Go back to take a look at figure 3.5; would you say this "lattice" is at the percolation threshold? Why?

2. This mathematical task is to show that in the regime of small p, the largest cluster on a 1-D lattice grows linearly with p; specifically,

$$\lim_{p \ll 1} S = \lim_{p \ll 1} \frac{1 + p}{1 - p} \simeq 1 + 2p. \qquad (4.9)$$

3. Construct a series of 1-D percolation lattices of length $N = 128$, with occupation probability ranging from $p = 0.1$ to $p = 0.9$ in steps of 0.1, like in figure 4.1. For each value of p, construct 10 such lattices, each using a different seed for the random number generator controlling the loading of the lattice (see section 4.2

and appendix C). Now, for each p value, determine the mean number of occupied nodes, as averaged over the 10 realizations of the lattice, and compare it to the expected value pN. Then, calculate the mean size of the largest cluster $\langle S \rangle$ for each p, again averaged over your 10 lattice realizations, and plot this mean value as a function of p. Identify the value of p at which the growth process switches from single-node addition to cluster fusion.

4. Generate a 2-D 256 × 256 Cartesian percolation lattice at $p = 0.59$, following the procedure described in section 4.2, and use the code listed in figure 4.3 (or some equivalent) to extract the largest cluster. Use the box-counting method introduced in the preceding chapter to determine its fractal index. Repeat the procedure at a few other values of p on either side of the percolation threshold, and verify that the fractal dimension of the largest cluster is smallest at $p = p_c$.

5. Generate a 10-member ensemble of 64 × 64 2-D Cartesian percolation lattices at $p = 0.59$, and build the cluster-size PDF for this data set, using logarithmically constant bin sizes, as described in appendix B.5. Estimate the power-law index by a linear least-squares fit to the logarithm of bin count versus logarithm of size. Now estimate the power-law index (and associated standard error) using the maximum likelihood estimator described in appendix B.6. How well do the two values compare?

6. And now the Grand Challenge! Percolation lattices can be used to study a phenomenon known as *anomalous diffusion*. The idea is as follows: First, generate a 2-D 512 × 512 lattice at its percolation threshold, identify the largest cluster, and place an ant-like agent (see section 2.4) on an occupied node near the center of this

cluster. At each temporal iteration, the ant selects randomly one of the four possible directions top/down/right/left, and steps to that location *only* if the node is occupied; otherwise the ant remains in place until the next temporal iteration. So, in essence, the ant is moving randomly in a "labyrinth" defined by the cluster of which the starting node is part. At each iteration n, calculate the (squared) displacement,

$$D_n^2 = (x_n - x_0)^2 + (y_n - y_0)^2,$$

from the ant's starting position (x_0, y_0). You may let the ant move over a preset number of time steps, but do stop the calculation if the ant reaches the edge of the lattice.

a. Repeat the above simulation process for 10 distinct realizations of your percolation lattice, and plot the ensemble-average rms distance $\sqrt{\langle D_n^2 \rangle}$ versus iteration count.

b. Repeat all of the above for lattices above and below the percolation threshold (at $p = 0.5$ to $p = 0.7$, say).

"Normal" diffusion is characterized by a displacement $\langle d_n \rangle \propto \sqrt{n}$ (see appendix C if needed). In which range of occupation probability p can diffusion be deemed most "anomalous"?

4.8 Further Reading

Much has been written on percolation as an exemplar of criticality. At this writing the classical reference remains

Stauffer, D., and Aharony, A., *Introduction to Percolation Theory*, 2nd ed., Taylor & Francis (1994);

but see also chapter 1 in

Christensen, K., and Moloney, N.R., *Complexity and Criticality*, London: Imperial College Press (2005).

The following offers a grand tour of phase transitions and related behaviors in a variety of physical, biological, and even social systems:

Solé, R.V., *Phase Transitions*, Princeton University Press (2011).

There is also much to be learned from the following book, for those with the appropriate mathematical skills:

Sornette, D., *Critical Phenomena in Natural Sciences*, Springer (2000).

Chapter 12 deals specifically with percolation, but the first four chapters also contain a wealth of useful information on critical systems and the statistical properties of variables distributed as power laws, or other distributions with power-law tails. Algorithms for cluster labeling exist, that are far more efficient than that introduced in section 4.2; see the Stauffer and Aharony book cited above, and also

Newman, M.J.E., and Ziff, R.M., "Efficient Monte Carlo algorithm and high-precision results for percolation," *Phys. Rev. Lett.*, **85**(19), 4104–4107 (2000);

as well as the following Santa Fe Institute working paper by the same authors (viewed June 2016):

http://www.santafe.edu/media/workingpapers/01-02-010.pdf.

On statistically proper techniques for assessing the probability of power-law behavior and determination of their indices from experimental data, see

Clauset, A., Shalizi, C.R., and Newman, M.E.J., "Power-law distributions in empirical data," *SIAM Review*, **51**(4), 661–703.

5

SANDPILES

The sky is blue, the sun is high, and you are sitting idle on a beach, a cold beer in one hand and a handful of dry sand in the other. Sand is slowly trickling through your fingers, and as a consequence, a small conical pile of sand is slowly growing below your hand. Sand avalanches of various sizes intermittently slide down the slope of the pile, which is growing both in width and in height *but maintains the same slope angle.*

However mundane this minor summer vacation event might appear, it has become the icon of *self-organized criticality* (hereafter, often abbreviated as SOC), an extremely robust mechanism for the autonomous development of complex, scale-invariant behaviors and patterns in natural systems. SOC will be encountered again and again in subsequent chapters, hiding under a variety of disguises, but here we shall first restrict ourselves to an extremely simple computational idealization of that iconic summertime pile of sand.

5.1 Model Definition

The sandpile model is a lattice-based CA-like system, evolving according to simple, discrete rules, local in space and time. Here we consider a 1-D lattice made up of N nodes with right+left neighbor connectivity, as in 1-D percolation (see figure 4.1). This lattice is used to discretize a real-valued variable S_j^n, where the subscript j identifies a node on the lattice and the superscript n denotes a temporal iteration. Initially ($n = 0$), we set

$$S_j^0 = 0, \quad j = 0, \ldots, N - 1. \tag{5.1}$$

This nodal variable is subjected to a forcing mechanism, whereby at each temporal iteration, a small increment s is added to the variable S, at a single, randomly selected node:

$$S_r^{n+1} = S_r^n + s, \quad r \in [0, N - 1], \quad s \in [0, \varepsilon], \tag{5.2}$$

where r and s are extracted from a uniform distribution of random deviates spanning the given ranges, and the maximum increment ε is an input parameter of the model. The physical system inspiring this simple model is a pile of sand, so you may imagine that S_j^n measures the height of the sandpile at the position j on the lattice at time n, and the forcing mechanism amounts to dropping sand grains at random locations on the pile. Obviously, the sandpile will grow in height in response to this forcing, at least at first.

Now for the dynamics of the system. As the pile grows, at each temporal iteration the magnitude of the *slope* associated with each nodal pair $(j, j + 1)$ is calculated:

$$z_j^n = |S_{j+1}^n - S_j^n|, \quad j = 0, \ldots, N - 2. \tag{5.3}$$

If this slope exceeds a preset critical threshold Z_c, then the nodal pair $(j, j + 1)$ is deemed unstable. This embodies the idea of static friction between sand grains in contact, which can equilibrate gravity up to a certain inclination angle, beyond which sand grains start toppling downslope. A redistribution rule capturing this toppling process is applied, so as to restore stability at the subsequent iteration. Here we use the following simple rule:

$$S_j^{n+1} = S_j^n + \tfrac{1}{2}(\bar{S} - S_j^n),$$
$$S_{j+1}^{n+1} = S_{j+1}^n + \tfrac{1}{2}(\bar{S} - S_{j+1}^n), \tag{5.4}$$

where

$$\bar{S} = \tfrac{1}{2}(S_{j+1}^n + S_j^n). \tag{5.5}$$

This rule displaces a quantity of sand, from the node with the higher S_j^n value to the other, such that the local slope z_j^n is reduced by a factor of 2. Figure 5.1 illustrates this redistribution process. If $\varepsilon \ll S_j, S_{j+1}$, then the critical slope is exceeded only by a small amount, and the above rule will always restore local stability. It is left as an easy exercise in algebra to verify that this rule is *conservative*, in the sense that sand is neither created nor destroyed by the redistribution:

$$S_j^{n+1} + S_{j+1}^{n+1} = S_j^n + S_{j+1}^n, \tag{5.6}$$

and that the quantity δS_j^n of sand displaced is given by

$$\delta S_j^n = \frac{z_j^n}{4}, \tag{5.7}$$

as indicated by the green boxes in figure 5.1. But now, even if the pair $(j, j+1)$ is the only unstable one on the lattice at iteration n, the redistribution has clearly changed the slope associated with the neighboring nodal pairs $(j-1, j)$ and $(j+1, j+2)$, since S_j^n and S_{j+1}^n have both changed; and it is certainly possible that one (or both) of these neighboring pairs now exceeds the critical threshold Z_c as a result. This is the case for the pair $(j+1, j+2)$ in the specific configuration depicted in figure 5.1. The redistribution rule is applied anew to that unstable nodal pair, but then the stability of its neighboring pairs must again be verified, and the redistribution rule applied once again if needed, and so on. This sequential process amounts to an *avalanche* of sand being displaced downslope, until every pair of contiguous nodes on the lattice is again stable with respect to equation (5.3).

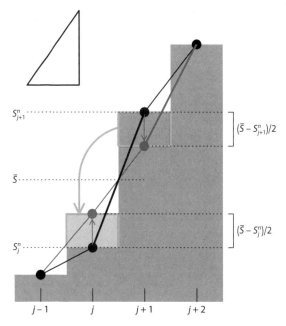

Figure 5.1. Action of the redistribution rules given by equations (5.4). The dark-gray columns indicate the nodal values (sand height) for a quartet of contiguous nodes, with the black solid dots linked by solid lines indicating the slope, as given by equation (5.3) and with thicker line segments flagging slopes in excess of the threshold Z_c (depicted by the triangular wedge at top left). Here, the nodal pair $(j, j + 1)$ exceeds this critical slope, so that the redistribution alters the nodal values as indicated by the two red vertical arrows. This is equivalent to moving, by one nodal spacing downslope, the quantity of sand enclosed by the upper green box, as indicated by the green arrow. This adjustment leads to the new slopes traced by the red dots and solid lines, which here is now unstable for the nodal pair $(j + 1, j + 2)$. This would lead to another readjustment at the next iteration (see text).

Now the boundary conditions come into play. At the last node of the lattice, at every iteration n we remove any sand having accumulated there due to an arriving avalanche:

$$S_{N-1}^n = 0. \tag{5.8}$$

This is as if the sandpile reached to the edge of a table, with sand simply falling off when moving beyond this position. No such removal takes place at the first node, which may be imagined as being due to the presence of a containing wall. The open boundary condition (5.8) turns out to play a crucial role here. Because the redistribution rule is conservative, and in view of the inexorable addition of sand to the system mediated by the forcing rule, the boundary is the only place where sand can be evacuated from the system.

In light of all this, one may imagine that a stationary state can be reached, characterized by a global slope equal to Z_c, with avalanches moving sand to the bottom of the pile at the same (average) rate as the forcing rule is loading the pile. As we shall see presently, a stationary state is indeed reached, but presents some characteristics one would have been very hard pressed to anticipate on the basis of the simple rules introduced above.

5.2 Numerical Implementation

The source code listed in figure 5.2 gives a minimal numerical implementation of our 1-D sandpile model, "minimal" in the sense that it favors coding clarity over computational efficiency and coding economy. Note the following:

1. The array sand[N] is our discrete variable S_j^n, and contains the quantity of sand at each of the N nodes of the lattice at a given iteration. Here, this is initially set to zero at all nodes (line 10).

2. The simulation is structured as one outer temporal loop, and this loop is set up to execute a predetermined number of temporal iterations n_iter (starting at line 14).

3. Each temporal iteration begins with an inner loop over each of the N-1 pairs of neighboring nodes on the lattice (starting on line 17). First, the local slope is

```
 1  # SLOPE-BASED SANDPILE MODEL IN ONE DIMENSION
 2  import numpy as np
 3  import matplotlib.pyplot as plt
 4  #-------------------------------------------------------------------------
 5  N=100                                   # Lattice size
 6  E=0.1                                   # Peak forcing increment
 7  critical_slope=5.                       # critical slope
 8  n_iter=200000                           # Number of temporal iterations
 9  #-------------------------------------------------------------------------
10  sand=np.zeros(N)                        # Lattice, initially empty
11  tsav=np.zeros(n_iter)                   # Avalanche time series
12  mass=np.zeros(n_iter)                   # Sandpile mass time series
13
14  for iterate in range(0,n_iter):         # Temporal iteration
15      move=np.zeros(N)                    # Initialize diplaced sand array
16
17      for j in range(0,N-1):             # Loop over lattice
18          slope=abs(sand[j+1]-sand[j])   # Eq (5.3): slope between j,j+1
19          if slope >= critical_slope:    # Pair j,j+1 is unstable
20              avrg=(sand[j]+sand[j+1])/2.
21              move[j]  +=(avrg-sand[j]  )/2. # Eq (5.4) sand moved to/from j
22              move[j+1]+=(avrg-sand[j+1])/2. # Eq (5.4) sand moved to/from j+1
23              tsav[iterate]+=slope/4.    # Eq (5.7) cumulate displaced mass
24      # end of lattice loop
25
26      if tsav[iterate] > 0:              # At least one node avalanched
27          sand+=move                     # Transfer sand
28      else:                              # No avalanche; drive lattice
29          j=np.random.random_integers(0,N-1) # Pick random node
30          sand[j]+=np.random.uniform(0,E)    # Eq (5.2): add sand increment
31
32      sand[N-1]=0.                       # Eq (5.8): boundary condition
33      mass[iterate]=np.sum(sand)         # Sandpile mass at this iteration
34  #   print("{0}, mass {1}.".format(iterate,mass[iterate]))
35  # End of temporal iteration
36
37  # Now plot a simpler version of Figure 5.4
38  plt.subplot(2,1,1)                      # Set up first plot (top)
39  plt.plot(range(0,n_iter),mass)          # Sandpile mass vs iteration
40  plt.ylabel('Sandpile mass')
41  plt.subplot(2,1,2)                      # Set up second plot (bottom)
42  plt.plot(range(0,n_iter),tsav)          # Displaced mass vs iteration
43  plt.ylabel('Displaced mass')
44  plt.xlabel('iteration')
45  plt.show()
46  # END
```

Figure 5.2. Source code in the Python programming language for the 1-D sandpile model described in the text. This represents a minimal implementation, emphasizing conceptual clarity over programming elegance, code length, or run-time speed.

calculated (line 18), then tested for stability (line 19), and wherever the stability criterion is violated, the quantity of sand that must be added or removed from each node to restore stability, as per the redistribution rule (5.4), is accumulated in the array move (lines 21–22), *without updating array* sand *at this stage*. This update is carried out only once all nodes have been tested, by adding the content of move to sand (line 27). This *synchronous update* of the nodal variable is important, otherwise a directional bias is introduced in the triggering and propagation of avalanches.

4. Addition of sand at a random node (lines 29–30) takes place only if the lattice was found to be everywhere stable at the current iteration. This is known as a "stop-and-go" sandpile, and is meant to reflect a *separation of timescale* between forcing and avalanching, the former being assumed to be a much slower process than the latter.

5. At the end of each iteration, the mass of the pile and the mass displaced by avalanches, to be defined shortly in equations (5.9) and (5.10), are stored in the arrays mass and tsav; these time series will be needed in the analyses to follow.

6. Note another piece of Python-specific coding on line 33: the instruction np.sum(sand), using the summing function from the NumPy library, returns the sum of all elements of array sand; this could be easily replaced by a loop sequentially summing the elements of the array.

5.3 A Representative Simulation

Let's look at what this code does for a small 100-node lattice, initially empty (i.e., $S_j^0 = 0$ for all j), with the driving amplitude

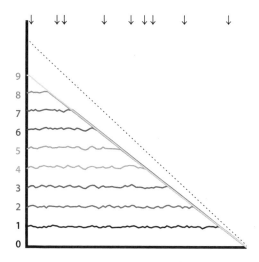

Figure 5.3. Growth of a 1-D sandpile constrained by a wall on its left edge, as produced by the code listed in figure 5.2, here starting from an empty $N = 100$ lattice, and with parameter values $Z_c = 5$ and $\varepsilon = 0.1$. The dotted line indicates a slope of Z_c. Each curve traces the top of the growing sandpile, and is separated from the preceding one by 10^5 iterations, as color coded from the bottom toward the top.

set at $\varepsilon = 0.1$ and the critical slope at $Z_c = 5$. Figure 5.3 illustrates the growth of the sandpile during the first 10^6 iterations. Recall that sand is being dropped at random locations on the lattice, but in a statistically uniform manner, so that at first the pile remains more or less flat as it grows. However, the "falloff" boundary condition imposed on the right edge drains sand from the pile, so that the pile develops a right-leaning slope, at first close to its right edge, but gradually extending farther and farther to the left. In contrast, at the left edge, the "wall" condition imposed there implies that sand just accumulates without falling off. Consequently the pile remains flat there until the slope growing from the right reaches the left edge. This occurs here after some 850,000 temporal iterations. In this *transient phase*, the system

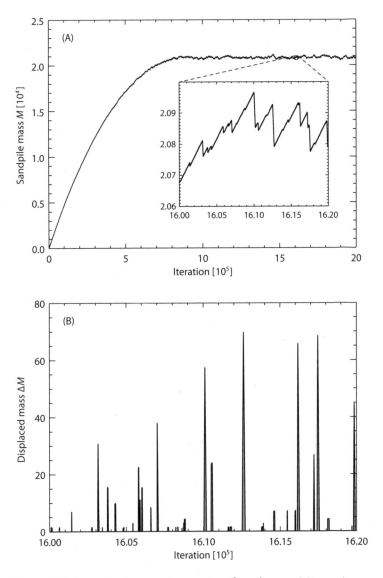

Figure 5.4. Panel A shows a time series of total mass M^n, as given by equation (5.9), for a simulation with parameter values $N = 100$, $Z_c = 5$, and $\varepsilon = 0.1$ and initial condition $S_j^0 = 0$. The inset shows a zoom of the time series in the statistically stationary phase of the

has not yet reached statistical equilibrium: averaged over many iterations, more sand is added to the pile than is evacuated at the open boundary.

This all makes sense and could have been easily expected, couldn't it, given the model's setup? So why bother to run the simulation? Well, to begin with, careful examination of figure 5.3 reveals that one very likely expectation did not materialize. The dotted line indicates the slope corresponding to the set critical slope $Z_c = 5$. In the statistically stationary state, the pile ends up with a slope significantly *smaller* (here by about 7%) than $Z_c = 5$. This equilibrium slope defines the *angle of repose* of the sandpile. But why does the pile stop growing *before* the critical slope is reached? This is due to the stochasticity embedded in the forcing mechanism, which leads to some nodal pairs going unstable before the pile as a whole has reached the critical slope Z_c. As a consequence, the system stabilizes at an average slope smaller than Z_c, approaching Z_c only in the limit $\varepsilon \rightarrow 0$. But this is just the beginning of the story.

It will prove useful to define a few global quantities in order to characterize the temporal evolution of the lattice. The most obvious is perhaps *mass*, namely, the total quantity of sand in the pile at iteration n:

$$M^n = \sum_{j=0}^{N-1} S_j^n. \tag{5.9}$$

Figure 5.4A shows a time series of this quantity, starting at the beginning of the simulation. Mass first grows with time during the transient phase, but eventually saturates at a value subjected

---

Figure 5.4 (*Continued*). simulation, highlighting its fractal shape. Panel B is a time series of displaced mass ΔM^n, as given by equation (5.10), spanning the same time interval as the inset on panel A. The matplotlib instructions in lines 38–45 of figure 5.2 produce similar plots.

to zero-mean fluctuations. These are better seen in the inset, which shows a zoom of a small portion of the time series. The shape is quite peculiar. In fact, the line defined by the M^n time series is self-similar, with a fractal dimension larger than unity. In this zoom, mass is seen to grow linearly, at a well-defined rate set by the magnitude of the forcing parameter ε, but this growth is episodically interrupted by sudden drops, occurring when sand is evacuated from the pile when avalanches reach the open boundary at the end of the lattice. The resulting fractal sawtooth pattern reflects the slow, statistically uniform loading and rapid, intermittent discharge. The sandpile is now in a *statistically stationary state*: the mass is ever varying, but its temporal average over a time span much larger than the mean time interval between two successive avalanches remains constant.

Another interesting quantity is the mass displaced at iteration n in the course of an ongoing avalanche:

$$\Delta M^n = \sum_{j=0}^{N-2} \delta S_j^n, \qquad (5.10)$$

where δS_j^n is given by equation (5.7). Keep in mind that this quantity is *not* necessarily equal to $M^{n+1} - M^n$, since an avalanche failing to reach the right edge of the sandpile will not lower the total mass of the pile, even though sand is being displaced downslope. Nonetheless, it is clear from figure 5.4 that the total mass of the sandpile varies very little, even when a large avalanche reaches the right boundary; the largest drop visible in the inset in figure 5.4A amounts to a mere 0.2% of the sandpile mass. This is because only a thin layer of sand along the slope is involved in the avalanching process, even for large avalanches. The underlying bulk of the sandpile remains "frozen" after the sandpile has reached its statistically stationary state.

Figure 5.4B shows the segment of the ΔM^n time series corresponding to the epoch plotted in the inset in part A.

This time series is again very intermittent, in the sense that $\Delta M^n = 0$, except during short "bursts" of activity corresponding to avalanches. These avalanches are triggered randomly, and have widely varying sizes, ranging from one pair of nodes, to the whole lattice.

Figure 5.5 illustrates the spatiotemporal unfolding of avalanches over 2000 iterations in the statistically stationary state of the same simulation as in figure 5.4. The vertically elongated images at the center and right each show a 1000-iteration segment, the right being a continuation of the central one, with time running vertically upward. The horizontal is the "spatial" dimension of the 1-D lattice, the open boundary being on the right. The square pixelated images on the left are two close-ups, each capturing the onset and early development of an avalanche. The color scale encodes the quantity of displaced sand, with green corresponding to zero. The purple/pink shades delineate the avalanching regions. Note how avalanches always start at a single nodal pair, following the addition of a sand increment at a single node, and typically expand downslope (here toward the right) as well as upslope (toward the left) in subsequent iterations. The smaller avalanches often remain contained within the slope (bottom of the middle image), but the larger ones typically reach all the way to the open boundary and discharge sand from the pile. The constant inclination angle of propagating avalanches in such diagrams reflects the one-node-per-iteration propagation speed of the avalanching front, as set by the local redistribution rule.

The aggressive color scale used to generate figure 5.5 was chosen to visually enhance substructures building up within avalanching regions. The most prominent pattern at the lattice scale is checkerboard-like, and simply reflects the fact that the stability and redistribution rules introduce a two-node spatial periodicity in the lattice readjustment. Of greater interest are the long-lived substructures emanating from the avalanching front

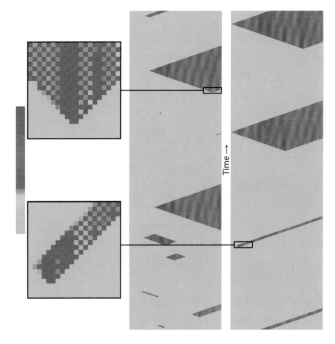

Figure 5.5. Spatiotemporal map of avalanches cascading across the lattice, in a 2000-iteration-long segment in the statistically stationary phase of the simulation plotted in figure 5.4. The image displays the displaced mass δS_j^n as a function of node number, running horizontally, and time, running vertically from bottom to top. The open boundary coincides with the right edge of each image. The image on the right is the temporal continuation of that in the middle, and the two pixelated images on the left are close-ups of the early phases of two avalanches. Green corresponds to zero displaced mass (stable slope), and shades of light blue through purple to red are avalanching regions. This rather unusual color scale was picked to better illustrate the substructures developing within avalanching regions (see text).

and propagating vertically upward in the avalanching regions. These are quite striking in the central and right images in figure 5.5. They are triggered by small variations in the slope characterizing stable regions in which the avalanching is progress-ing. These irregularities are responsible for avalanches, even large ones, sometimes stopping prior to reaching one or the other lattice boundaries. Morphologically, they also bear some simi-larity to the spatiotemporal structures that can build up in 1-D two-state CAs of the type investigated in section 2.1.

5.4 Measuring Avalanches

Figures 5.4B and 5.5 illustrate well the disparities in avalanche size and shape. This is worth looking into in greater detail. We begin by defining three global quantities characterizing each avalanche, all computable from the time series of displaced sand (array `tsav` in the simulation code listed in figure 5.2):

1. avalanche energy[1] E: the sum of all displaced mass ΔM^n over the duration of a given avalanche;
2. avalanche peak P: the largest ΔM^n value produced in the course of the avalanche;
3. avalanche duration T: the number of iterations elapsed between the triggering of an avalanche and the last local redistribution that follows.

These three quantities can be easily extracted from the time series of displaced mass (array `tsav` in the Python code listed in figure 5.2). The idea is to identify the beginning of an avalanche as a time step `iterate` for which `tsav(iterate)>0` but `tsav(iterate-1)=0`; likewise, an avalanche ends at iteration `iterate-1` if `tsav(iterate-1)>0` but `tsav(iterate)=0`.

---

[1]"Energy" is used here somewhat loosely, yet clearly the redistribution rules involve displacing sand downslope, as indicated by the green boxes in figure 5.1, thus liberating gravitational potential energy, and justifying the analogy.

The following user-defined Python function shows how to code this up:

```
1   # FUNCTION MEASURE_AV: EXTRACTS ENERGY, PEAK, AND DURATION OF AVALANCHES
2   def measure_av(n_iter,tsav):
3       n_max_av=10000                          # max number of avalanches
4       e_av=np.zeros(n_max_av)                 # avalanche energy series
5       p_av=np.zeros(n_max_av)                 # avalanche peak series
6       t_av=np.zeros(n_max_av)                 # avalanche duration series
7       n_av,sum,istart,avmax=-1,0,0,0.
8       for iterate in range(1,n_iter):         # loop over time series
9           if tsav[iterate] > 0. and tsav[iterate-1] == 0.:
10              sum,avmax=0.,0.
11              istart=iterate                  # a new avalanche begins
12              if n_av == n_max_av-1:          # safety test
13                  print("too many avalanches")
14                  break                       # break out of loop
15              n_av+=1                         # increment counter
16          sum+=tsav[iterate]                  # cumulate displaced mass
17          if tsav[iterate] > avmax:           # check for peak
18              avmax=tsav[iterate]
19          if tsav[iterate] <= 0. and tsav[iterate-1] > 0: # avalanche ends
20              e_av[n_av]= sum                 # avalanche energy
21              p_av[n_av]= avmax               # avalanche peak
22              t_av[n_av]= iterate-istart      # avalanche duration
23
24      # end of loop over time series
25      return n_av,e_av,p_av,t_av
26  # END FUNCTION MEASURE_AV
```

This function could be called, for example, after the outer loop in the sandpile code of figure 5.2. Note the safety test (lines 12–14) exiting the loop so as to avoid the avalanche counter n_av becoming larger than n_max_av, which would cause out-of-bounds indexing of the arrays e_av, p_av, and t_av. Upon exiting from the loop, the variable n_av contains the number of avalanches in the time series array tsav, and the arrays e_av, p_av, and t_av contain the associated energy E, peak displaced mass P, and duration T of each of these avalanches.

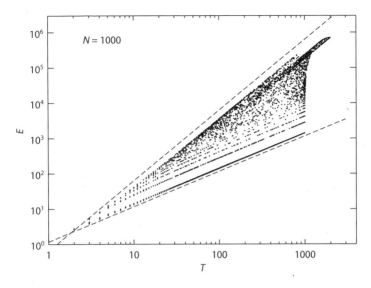

Figure 5.6. Correlation between avalanche size E (displaced mass) and duration T in the statistically stationary phase of a sandpile simulation on an $N = 1000$ 1-D lattice. The dotted lines bracketing the avalanche data have slopes of $+1$ and $+2$ in this log-log plot, corresponding respectively to the relationships $E \propto T$ and $E \propto T^2$.

Although large avalanches moving more sand tend to last longer and reach higher peak discharge rates, the quantities E, P, and T are correlated only in a statistical sense. Figure 5.6 shows the correlation between avalanche size E and duration T for 15,019 avalanches having occurred in a 5×10^6 iteration segment of a simulation on an $N = 1000$ lattice. Overall E does increase with T, but the distribution of avalanche data shows some rather peculiar groupings, most notably along diagonal lines in this correlation plot. Moreover, all data fall within a wedge delimited by lines with slopes of $+1$ and $+2$ in this log-log plot.

Consider a lattice everywhere at the angle of repose, with the addition of a small random increment at node j bringing one nodal pair infinitesimally beyond the stability threshold. Equation (5.7) then yields a displaced mass $\delta S_j^n = Z_c/4$; this is the smallest avalanche that can be produced on the lattice; it is the "quantum" of displaced mass (or energy) for this system, hereafter denoted δM_0. Now, suppose that this redistribution destabilizes the downslope pair $(j, j + 1)$, but not its upslope counterpart $(j - 1, j)$; with the lattice everywhere at the angle of repose, our quantum of displaced mass will move down the slope, one node per iteration, until it is evacuated at the open boundary. If the original unstable nodal pair is M nodes away from the open boundary, this avalanche will have duration $T = M$ and energy $E = M \times \delta M_0$; consequently, $E = \delta M_0 T$, a linear relationship. If the initial avalanche destabilizes both neighboring pairs but no other pair upslope, then 2 quanta of mass will move down the slope, leading to $E = 2\delta M_0 T$, and so on for higher numbers of mass quanta. The duration of such avalanches is clearly bounded by the size of the lattice. These are the line-like avalanches in figure 5.5, and they map onto the straight-line groupings with slope $+1$ in figure 5.6. The avalanche whose onset is plotted in the bottom-left close-up in figure 5.5 belongs to the fourth such family (4 mass quanta moving out to the open boundary). These families represent the quantized "energy levels" accessible to the avalanches. The upper bounding line with slope of $+2$ is associated with avalanches spreading both upslope and downslope; all nodes in between avalanche repeatedly until stabilization occurs at the ends of the avalanche front, or mass is evacuated at the boundary. These are the avalanches taking the form of solid wedges in figure 5.5. In such cases, the number of avalanching nodes increases linearly with T, so that the time-integrated displaced mass will be $\propto T^2$. The locality of the redistribution rules precludes avalanches from growing faster on this 1-D lattice, which then explains why the avalanche energies are bounded from above by a straight line of slope $+2$ in figure 5.6. Of course, any intermediate avalanche shape between lines and wedges is possible, and so the space between the two straight lines is also populated by the avalanche data. Incidentally, there is a lesson lurking here: just because a system is deemed to exhibit "complexity" does not mean that some aspects of its global behavior cannot be understood straightforwardly!

Even though the correlations between avalanche parameters exhibit odd structure, their individual statistical distributions are noteworthy. Figures 5.7A,B show the PDFs (see appendix C) for E and P, for simulations carried out over lattices of sizes $N = 100, 300, 1000$, and 3000, but otherwise identical ($Z_c = 5$, $\varepsilon = 0.1$, and redistribution given by equation (5.4)). The PDFs take the form of power laws, with logarithmic slope independent of lattice size; as the latter increases, the distribution simply extends farther to the right.

This is behavior we have encountered before, in chapter 4, in the size distribution of clusters on 2-D lattices at the percolation threshold (cf. figure 4.8). Here this invariant power-law behavior materializes only in the statistically stationary phase of the simulation. It indicates that avalanches are self-similar, i.e., they do not have a characteristic size. This scale invariance reflects the fact that at the dynamical level, the only thing distinguishing a large avalanche from a small one is the number of lattices nodes involved; the same local rules govern the interaction between nodes. But in the percolation context, we also argued that scale invariance appeared only when the system had reach a critical state; could this also be the case here?

5.5 Self-Organized Criticality

It is truly remarkable that of all the possible ways to move sand downslope at the same average rate as sand addition by the forcing rule, so as to achieve a statistically stationary state, our sandpile model "selects" the one characterized by scale-free avalanches.

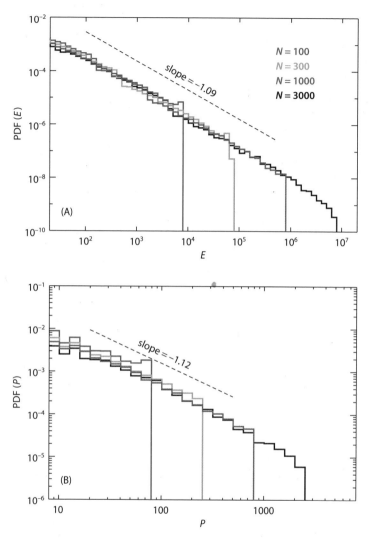

Figure 5.7. Probability density function of (A) avalanche energy E and (B) avalanche peak P, in the statistically stationary states of the sandpile model for varying lattice sizes, as indicated. The PDF of avalanche duration T resembles that for P in panel B, except for a steeper logarithmic slope. Note the logarithmic scales on both axes. In all cases, the PDFs take the form of power laws, with a

Because many natural systems behave in this manner, the sandpile (real or idealized) has become the icon for avalanching behavior in general, and for the concept of *self-organized criticality* in particular.

We saw in chapter 4, in the context of percolation, that a system is deemed critical when the impact of a small, localized perturbation can be felt across the whole system. Recall how, at the percolation threshold, occupying one more node on the lattice can connect two preexisting clusters, forming a single large cluster spanning the whole lattice; as a result the system suddenly becomes permeable, electrically conducting, whatever, whereas prior to that it was impermeable, or insulating, etc. You should also recall that this extreme sensitivity materialized only at the percolation threshold, so that critical behavior required external fine-tuning of a control parameter, which in the case of percolation is the occupation probability p. Moreover, it is only at the percolation threshold that clusters on the lattice exhibited scale invariance (see figure 4.7).

So where is the criticality here? With the sandpile, the equivalent of the percolation threshold is the angle of repose of the pile. If the slope is inferior to this, as when the sandpile is still growing, then local addition of sand may trigger small, spatially confined avalanches, but certainly nothing spanning the whole lattice. If the global slope angle is larger than the angle of repose, then the lattice is already avalanching vigorously. Only at the angle of repose can the addition of a small bit of sand at a single random node do anything between (1) nothing,

---

Figure 5.7 (*Continued*). flattening at small values of E and P, and a sharp drop at high values, occurring at progressively larger values of E and P for larger lattices. Note, however, that for large enough lattices the logarithmic slope is independent of lattice size. Compare this to figure 4.8.

and (2) trigger an avalanche running along the whole slope. However, unlike with percolation, here the angle of repose is reached "naturally," as a consequence of the dynamical evolution of the system—namely, the forcing, stability, and redistribution rules—through interactions between a large number of lattice nodes over time, without any fine-tuning of external parameters. The critical state is here an *attractor* of the dynamics. For this reason, systems such as the sandpile are said to be in a state of *self-organized* criticality, to distinguish them from conventional critical systems which rely on external fine-tuning of a control parameter.

Much effort has gone into identifying the conditions under which a system can exhibit self-organized critical behavior. At this writing, there exists no general theory of self-organized critical systems, but the following characteristics appear sufficient—and possibly even necessary. A system must be

1. open and dissipative;
2. loaded by slow forcing;
3. subjected to a local threshold instability...
4. ...that restores stability through local readjustment.

However restrictive this may appear, the number and variety of natural systems that, in principle, meet these requirements is actually quite large. Joining avalanches and other forms of landslides are forest fires, earthquakes, hydrological drainage networks, geomagnetic substorms, and solar flares, to mention but a few. Some of these we will actually encounter in subsequent chapters. More speculative applications of the theory have also been made to species extinction and evolution by punctuated equilibrium, fluctuations and crashes of stock markets, electrical blackouts on power grids, and wars. For more on these topics, see the references listed in the bibliography at the end of this chapter and of chapter 9.

5.6 Exercises and Further Computational Explorations

1. Verify that the redistribution rule given by equation (5.4) does lead to equation (5.7).

2. Modify the 1-D sandpile simulation code of figure 5.2 to keep track of the mass falling off the pile at its right edge. This will be an avalanching time series distinct from the displaced mass time series `tsav`. Once the statistically stationary state has been reached, use this new "falloff" time series to calculate the corresponding avalanche parameters E, P, and T, as in section 5.4 above, and construct the corresponding PDFs (as in figure 5.7). Are falloff avalanches scale invariant? How well does the "falloff E" correlate with the "avalanching E" as defined in section 5.4?

3. Use the 1-D sandpile simulation code of figure 5.2 to verify that the statistically stationary, self-organized critical state is independent of the initial condition; more specifically, try various types of initial conditions such as, for example, an initial sandpile at the angle Z_c, or already at the angle of repose, or an initial sandpile loaded uniformly at some fixed height, etc.

4. Carry out 100-node simulations using different ε (0.01, 0.1, and 1, say). Are the angles of repose the same? Making sure to have reached the statistically stationary state before beginning your analyses, construct a PDF of slope values (as given by equation (5.3)) as extracted from a single non-avalanching iteration of each simulation; are these PDFs dependent on the value of ε? Then construct the PDF of avalanche energy E for the same three simulations; are they the same?

5. The 1-D sandpile code listed in figure 5.2 is very inefficient from the computational point of view; most notably perhaps, at every iteration it checks *all* lattice

nodes for stability, even if a perturbation s has been added only at a single, randomly selected node at the preceding iteration (see equation (5.2)). An easy way to improve on this is to modify the start and end points of the loop over the lattice nodes so that stability is checked only at the three nodes $[r - 1, r, r + 1]$, where r is the random node at which a perturbation is added. The reader with prior coding experience may instead try the really efficient algorithmic approach, which is to keep a *list* of nodes either avalanching or subject to forcing, and run the stability checks and redistribution operations only on list members and their immediate neighbors. This is fairly straightforward in Python, which contains a number of computationally efficient list manipulation operators and functions. This may sound like a lot of work to speed up a simulation code, but when generalizing the avalanche model to two or three (or more) spatial dimensions, such a "trick" will mean waiting 10 minutes for the simulation to run, rather than 10 hours (or more).

6. The Grand Challenge for this chapter is to design a 2-D version of the sandpile model introduced herein. Your primary challenge is to generalize the stability criterion (equation (5.3)) and redistribution rule (equation (5.4)) to 2-D. Begin by thinking how to define the slope to be associated with a 2×2 block of nodes. Measure the avalanche characteristics E, P, and T once the SOC state has been reached, and verify that these are distributed again as power laws. Are their indices the same as in the 1-D case? You should seriously consider implementing in your 2-D sandpile code at least the first of the speedup strategies outlined in the preceding exercise.

5.7 Further Reading

The concept of SOC was created by Per Bak, who became its most enthusiastic advocate as a theory of (almost) everything. His writings on the topic are required reading:

Bak, P., Tang, C., and Wiesenfeld, K., "Self-organized criticality: An explanation of the $1/f$ noise," *Phys. Rev. Lett.*, **59**, 381 (1987);
Bak, P., *How Nature Works*, New York: Springer/Copernicus (1996);

but also see

Jensen, H.J., *Self-Organized Criticality*, Cambridge University Press (1998);

and, at a more technical level,

Turcotte, D.L., "Self-organized criticality," *Rep. Prog. Phys.*, **62**(10), 1377–1429 (1999);
Sornette, D., *Critical Phenomena in Natural Sciences*, Berlin: Springer (2000);
Hergarten, S., *Self-Organized Criticality in Earth Systems*, Berlin: Springer (2002);
Aschwanden, M.J., ed., *Self-Organized Criticality Systems*, Berlin: Open Academic Press (2013).

Finally, for a good reality check on the behavior of real piles of real sand,

Duran, J., *Sands, Powders, and Grains*, New York: Springer (2000).

It turns out that real piles of real sand seldom exhibit the SOC behavior characterizing the idealized sandpile models of the type considered in this chapter. However, some granular materials do, including rice grains; see chapter 3 in the book by Jensen listed above.

6

FOREST FIRES

Chapter 4 introduced some of the remarkable properties of randomly produced percolation clusters. These clusters were entirely static, "frozen" objects, their structure determined once and for all by the specific realization of random deviates used to fill the lattice.

Can any "natural" process generate dynamically something conceptually resembling a percolation cluster? The answer is yes, as exemplified by the forest-fire model investigated in this chapter. Its ecological inspiration is probably as far removed as it could be from flow through porous media or phase transitions, yet at a deeper level it does represent an instance of dynamical percolation.

6.1 Model Definition

The forest-fire model is, fundamentally, a probabilistic CA. Sticking again to a 2-D Cartesian lattice, each node (i, j) is assigned a state $s_{i,j}$ which can take one of three possible values: "empty," "inactive," and "active." Starting from an empty lattice $(s_{i,j} = 0$ for all i, $j)$, the nodal variable evolves in discrete time steps ($s_{i,j}^{n} \rightarrow s_{i,j}^{n+1}$) according to a set of local rules, some of a stochastic nature:

Rule 1. An empty node can become occupied with probability p_g (stochastic).

Rule 2. An inactive node can be activated with probability p_f (stochastic).

Rule 3. An inactive node becomes active if one or more of its nearest neighbors was active at the preceding iteration (deterministic).

Rule 4. Active nodes become empty at the following iteration (deterministic).

The ecological inspiration of the model should be obvious: inactive nodes represent trees; active nodes are burning trees; Rule 1 is tree growth; Rule 2 is a tree being ignited by lightning; Rule 3 is fire jumping from one tree to a neighboring tree; and Rule 4 is destruction of a tree by fire. You have probably anticipated already that successive ignition of trees by a burning neighbor can lead to the propagation of a burning "front" across the lattice, i.e., an "avalanche" of burning trees. This expectation is certainly borne out, but as with the simple sandpile model considered in the preceding chapter, the spatiotemporal evolution of the system holds quite a few surprises in store for us.

6.2 Numerical Implementation

The Python source code listed in figure 6.1 is a minimal implementation of the forest-fire model, again in the sense that it sacrifices coding conciseness and execution speed to conceptual clarity and readability. The overall structure is similar to the sandpile code of figure 5.2, but the simulation is now performed on a 2-D Cartesian lattice of size $N \times N$. Burning (active) nodes are assigned the numerical value 2, while occupied (inactive) nodes are set to 1, and empty nodes to 0. The temporal iteration is governed by the outer fixed-length loop starting on line 15, inside

```
1   # FOREST-FIRE MODEL ON 2D CARTESIAN LATTICE
2   import numpy as np
3   import matplotlib.pyplot as plt
4   #-------------------------------------------------------------------------
5   N     =100                               # Lattice size
6   p_g   =1.e-3                             # Growth probability
7   p_f   =1.e-5                             # Lightning probability
8   n_iter=25000                            # Number of temporal iterations
9   #-------------------------------------------------------------------------
10  dx=np.array([-1,0,1,1,1,0,-1,-1])        # Template arrays
11  dy=np.array([-1,-1,-1,0,1,1,1,0])
12  grid=np.zeros([N+2,N+2],dtype='int')        # Initialize lattice: no trees
13  trees=0                                  # Tree counter
14
15  for iterate in range(0,n_iter):          # temporal iteration
16      update=np.zeros([N+2,N+2],dtype='int')   # evolution array
17      burn=0                               # burning tree counter
18      # scan lattice to flag which trees must grow, ignite or vanish
19      for i in range(1,N+1):
20          for j in range(1,N+1):
21              if grid[i,j] == 1:               # there is a tree on this node
22                  if 2 in grid[i+dx[:],j+dy[:]]: # 1 or more burning neighbor
23                      update[i,j]=1              # ignite
24                      burn+=1
25                  if np.random.uniform() < p_f: # lightning strikes (maybe)
26                      update[i,j]=1              # ignite
27                      burn+=1
28              if grid[i,j] == 2:               # remove trees already burning
29                  update[i,j]=-2
30                  trees-=1
31              if grid[i,j] == 0:               # empty node
32                  if np.random.uniform() < p_g: # grow tree (maybe)
33                      update[i,j]=1
34                      trees+=1
35      # end of lattice scan
36
37      grid+=update                         # synchronous update lattice
38      print("iteration {0}, trees {1}, burn {2}.".format(iterate,trees,burn))
39  # end of temporal loop
40  plt.imshow(grid,interpolation="nearest")    # display final state
41  plt.show()
42  # END
```

Figure 6.1. A minimal implementation of the forest-fire model in the Python programming language.

which really all the coding action is sitting. Take note of the following:

1. The simulation begins with an empty lattice: all nodal values in grid are set to 0 (line 12), using Python/NumPy's array creation-and-initialization function zeros.

2. As with the DLA code of chapter 3, the 2-D arrays
   grid and update are padded with an outer frame of
   ghost nodes which always remain empty, but allow
   nodes at the real edges of the lattice to be tested for
   ignition in the same manner as interior nodes.
   Consequently, even though the lattice array is of size
   $(N+2) \times (N+2)$, loops over the lattice run from 1 to N,
   meaning in Python range(1,N+1) on lines 19–20, as
   per the loop range and array element numbering
   conventions in the Python programming language. See
   appendix D.1 if needed.

3. Ghost nodes retain the value 0 throughout the
   simulation; you may think of this as equivalent to the
   simulation domain being enclosed within four
   scrupulously well-maintained fire trenches.

4. Once again the nodes that are to grow a tree, catch fire,
   or become empty are first identified in the first block of
   for loops, and the needed changes (+1 for tree growth,
   line 33; +1 for igniting an existing tree, either by
   lightning (line 23) or via a burning neighbor (line 26);
   and -2 for a burned tree vanishing (line 29)) are stored
   in the 2-D array update. This work array is reset to
   zero at the beginning of each temporal iteration (line
   16).

5. The lattice update is later carried out synchronously, at
   the end of the temporal iteration loop (line 37).

6. The relative coordinates of the 8 nearest neighbors to
   any node are stored in the template arrays dx and dy
   (lines 10–11). See appendix D.1 if needed.

7. Again note, on line 22, the Python-specific instruction

   if 2 in grid[i+dx[:],j+dy[:]]:

   and its built-in implicit loop, to check whether there is
   at least one burning tree in the set of nearest neighbors

to node (i, j), as defined by the template arrays
dx and dy.

8. Note that lightning can still strike while a fire is
burning; this model is operating in "running" rather
than "stop-and-go" mode.

In case you have not noticed it already, this forest-fire model
is at the core of the algorithm introduced in chapter 4 for
the tagging of clusters on the percolation lattice (if you are
not convinced, compare figures 4.3 and 6.1). Occupied nodes
are the trees; tree growth is turned off, and random ignition by lightning strikes is replaced by systematic ignition of
as-yet-untagged occupied nodes. The ensemble of trees burned
by each such ignition is a cluster, and the largest fire maps are the
largest cluster on the lattice.

Getting back to the forest-fire model per se, clearly the rules
governing the lattice evolution are quite simple, and only Rule 3
actually involves nearest-neighbor contact. Moreover, the model
involves only two free parameters, namely, the tree growth
probability p_g and the lightning probability p_f. Nonetheless,
as these two parameters are varied, the model can generate a
surprisingly wide range of behaviors, hard to anticipate on the
basis of its defining dynamical rules.

6.3 A Representative Simulation

Figure 6.2 shows the triggering, growth, and decay of a large fire
in a representative forest-fire model simulation on a small 100 ×
100 lattice, with parameter values $p_g = 10^{-3}$ and $p_f = 10^{-5}$.
This simulation had already been running for many thousands of
iterations, and so had reached a statistically stationary state.[1] The

---

[1] In this forest-fire model this is best ascertained by tracking the total number
of trees on the lattice, until it levels off to a stable mean value. Note also
that whatever the initial conditions, the duration of the initial transient phase
increases rapidly with decreasing p_g and p_f.

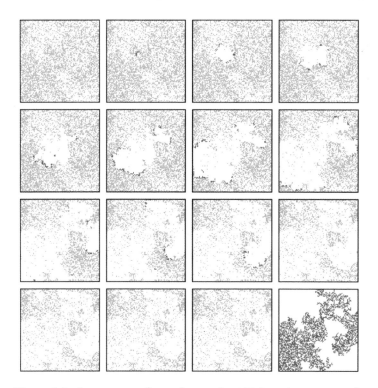

Figure 6.2. A sequence of snapshots, taken 10 iterations apart, of a 100×100 lattice in a simulation of the forest-fire model running with $p_g = 10^{-3}$ and $p_f = 10^{-5}$. Empty nodes are left white, nodes occupied but inactive are green, and active nodes are red. Here lightning has struck a bit left and up of the lattice center, two iterations prior to the second snapshot. The resulting burning front subsequently sweeps through a large fraction of the lattice. The bottom-right frame shows the location of all trees having burned in this fire. Notice also the small fire, triggered by a second lightning strike, ignited in the upper-left portion of the lattice a few iterations prior to the eleventh frame (third column in third row from top).

figure shows a sequence of snapshots taken 10 iterations apart, going left to right from top to bottom. A few iterations prior to the second snapshot, lightning has struck a bit up and left from the lattice center. Fire activity propagates from node to node, at an average speed determined by the density of trees but never exceeding one lattice spacing (horizontally and/or vertically) per iteration, as per Rule 3. The combustion front is initially almost circular, but later evolves into a far more convoluted shape as the fire sweeps across the lattice, reflecting the substantial spatial variations of tree density in the pre-fire lattice configuration. This heterogeneity is itself a consequence of previous fires having burned across the lattice in the more or less distant past (see the last snapshot in figure 6.2). Even though fires are triggered by a stochastic process (Rule 2 above), past fire activity influences the evolution of current fires.

Figure 6.2 illustrates well the disparity of timescales characterizing the forest-fire model. The shortest is the "dynamical" timescale characterizing the propagation of the fire from one node to a neighboring node, namely, one temporal iteration. The next timescale is that associated with tree growth, and is given by $p_{\mathrm{g}}^{-1} = 10^3$ iterations here. Starting with an empty 100×100 lattice, this means that, on average, 10 new trees would grow at each temporal iteration. The first three snapshots in the last row exemplify quite well how much longer than the dynamical timescale this is; they must be scrutinized very carefully to notice the $\simeq 300$ new trees that have appeared in the course of the 30 iterations spanned by these snapshots. The spontaneous activation probability—lightning strikes—usually determines the longest timescale. The expected time interval between two successive activations is of the order of $(p \times N^2 \times p_{\mathrm{f}})^{-1}$, where N is the linear size of the lattice and p the mean occupation probability in the statistically stationary state; this is defined as the number of live trees divided by N^2, the total number of lattice nodes.[2]

---

[2]Here, the occupation probability p is not an input parameter, as in percolation, but is a characteristic of the statistically stationary state attained by the simulation, but even then, only to a first approximation (more on this shortly).

Here, with $N = 100$, $p \sim 0.2$, and $p_f = 10^{-5}$, a lightning strike is expected every 50 iterations on average, but it must be kept in mind that, following a large fire such as in figure 6.2, p can fall much below its mean value calculated over the duration of the simulation.

The bottom-right panel in figure 6.2 shows the "cluster" of all trees burned in the 105-iteration-long fire covered by the other frames. Overall, this maps well, but not perfectly, to the tree density characterizing the top-left panel of figure 6.2, just prior to fire onset. Note how this cluster of burned trees contains "holes" in which clumps of trees have survived the fire, as the burning front became more convoluted. The shape of this cluster should remind you of the percolation clusters encountered in chapter 4; and yes, you hopefully guessed it, this cluster of burned trees is a fractal.

Figure 6.3 shows a segment of the time series of burning trees, in the same simulation. The large fire of figure 6.2 is the largest of the three fires visible on this time series, starting at iteration 3055. Fires clearly span a wide range in size, and their activity can show significant temporal variability in the course of a given fire. As one might have expected, large fires destroying large numbers of trees tend to burn longer and flare up more strongly, but the correlation between these fire measures is far from perfect; in figure 6.3, the third fire lasts only a few iterations more than the second, but destroys almost three times as many trees.

6.4 Model Behavior

The numerical choices made for the growth and activation probabilities p_g and p_f can lead to widely varying behaviors in the spatiotemporal evolution of the system. This is illustrated in figure 6.4 (and in figure 6.6 discussed later), which shows time series of the number of occupied nodes that are inactive, N_a (i.e., occupied by a tree; in green) and active, N_f (burning trees; in red), for four different combinations of p_g and p_f values.

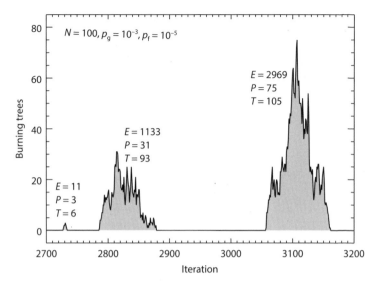

Figure 6.3. Time series of the number of burning trees in the simulation of figure 6.2, with the large fire starting at iteration 3055 being the one captured by that sequence of snapshots. Next to each fire are listed the total number of burned trees (E), the peak number of burned trees at any single iteration (P), and the fire duration (T).

If $p_f \sim p_g$ (top panel of figure 6.4, with $p_g = p_f = 10^{-4}$), then trees are struck by lightning at· a frequency comparable to their growth rate. The total number of trees remains approximately constant, and numerous small fires are always burning here and there, without ever becoming large, because the density of trees is too small; with N_a hovering around 1500, only 15% of the 10^4 lattice nodes are occupied at any given time, meaning that few pairs of trees stand on neighboring nodes. If p_g is raised to 10^{-2} (bottom panel in figure 6.4), trees grow much faster and their density is roughly twice as much. Not only can fires now spread, but in fact trees are now growing so rapidly that once ignited, a fire never stops because new growth behind the

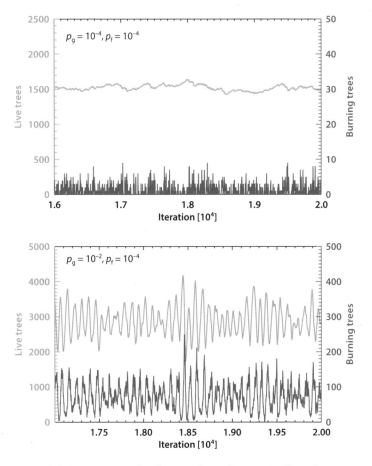

Figure 6.4. Time series for the number of active (red) and inactive (green) occupied nodes, for various combinations of p_g and p_f, in a regime where these growth and activation probabilities are relatively high. Both of these simulations are run on a 100×100 lattice, and the time series plotted are extracted far into the statistically stationary state.

burning front replenishes the forest at a rate comparable to the time it takes the fire to move across the lattice, here of the order of 100 iterations.

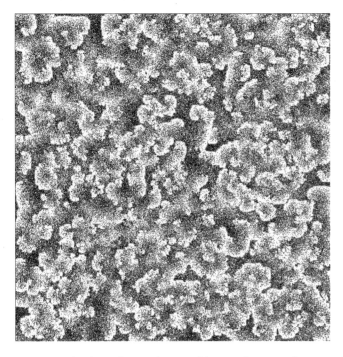

Figure 6.5. Snapshot of a 1024 × 1024 lattice, for a simulation with $p_g = 10^{-2}$ and $p_f = 10^{-5}$, but with lightning artificially turned off after the first 50 iterations. In this parameter regime, trees grow so fast that once ignited, fire persists throughout the simulation, with burning fronts expanding, fragmenting, shrinking, and interacting with one another. Note how the curved burning fronts often show a tendency to spiral inward at their extremities.

These parameter regimes are of course ecologically unrealistic, but represent classes of possible behavior for this model that are quite interesting in their own right. Figure 6.5 shows a snapshot of a 1024 × 1024 lattice in the rapid regrowth regime ($p_g = 10^{-2}$), where a few dozen random lightning strikes have taken place in the first 50 iterations, but lightning has been artificially "turned off" afterward. Moving burning fronts (in red) are ubiquitous across the lattice, growing, shrinking, fragmenting, merging, and interacting with one another, and often develop into approximately circular arcs, with their tips curling back inward, in the manner of a spiral with a large opening angle.[3] The density of trees (in green) at any location on the lattice undergoes a recurrence cycle of slow growth at a rate set by p_g, up to a value approaching unity, then a sudden drop to zero as the burning front moves through, followed by slow growth anew. This general type of recurrence cycle will be encountered repeatedly in subsequent chapters.

Let's get back to the more ecologically realistic situation where trees grow slowly and fires are rare events. Figure 6.6 shows what happens if the growth and activation probabilities are lowered to the much smaller values $p_g = 10^{-3}$ and $p_f = 10^{-6}$. The lattice now has enough time to really fill up before lightning strikes again. But when the ignition finally happens, almost every tree has at least one nearest neighbor, so the fire sweeps almost the whole lattice clean. This leads to a quasiperiodic "load/unload" recurrence cycle whereby, at more or less regular intervals, the whole forest is destroyed, and regrowth must start from zero or nearly so. In the top panel of figure 6.6, when lightning strikes there are around 5000 occupied nodes, out of a possible grand total of $100 \times 100 = 10,000$. The corresponding occupation probability is therefore $\simeq 0.5$, which pretty much guarantees that every tree has a neighbor. The fact that a few hundred trees remain at the end of a large fire is in part a boundary effect; on a 100×100 lattice, 392 nodes are boundary nodes that have three fewer neighbors than interior nodes, and the 4 corner nodes have even fewer. These boundary nodes are thus harder to reach for an ongoing fire.

At low p_f, the only way to break the load/unload cycle so prominent in the top panel of figure 6.6 is if tree growth is

---

[3]In chapter 11, we will encounter a CA behaving similarly, when discussing excitable systems and reaction–diffusion chemical reactions.

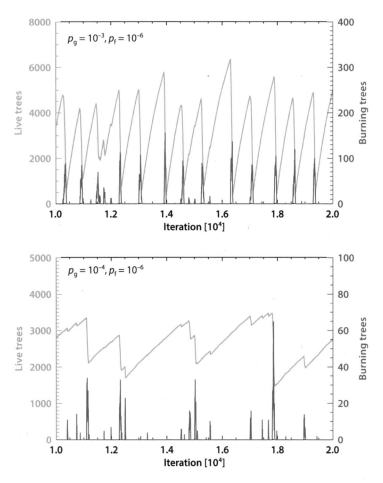

Figure 6.6. Identical in format to figure 6.4, but now for simulations operating in the regime where the activation probability p_f is very small. Compare the bottom plot to the inset in figure 5.4A.

sufficiently slow that the lattice does not have time to completely fill up between two successive lightning strikes. Keeping $p_f = 10^{-6}$ but lowering p_g to 10^{-4} yields the solution plotted in the bottom panel of figure 6.6. Fires, when they occur, can still be quite large, but they are now triggered far less regularly and exhibit a wide range of sizes. Note also the fractal sawtooth pattern of the time series for occupied nodes, which shows an uncanny resemblance to the mass time series in the sandpile model of the preceding chapter (cf. the inset in figure 5.4A).

In cases like in figure 6.6, where one or more fires are not burning continuously somewhere on the lattice (as they do in figure 6.5), it is possible to characterize each individual fire as we did for avalanches in the sandpile model, through the variables E, P, and T, defined respectively as

1. E: the total number of trees burned in the fire;
2. P: the peak number of trees burned at any one iteration in the course of the fire;
3. T: the fire duration, measured in temporal iterations.

These quantities are correlated with one another, in that large fires tend to last longer, but we know already from figure 6.3 that a perfect correlation is not to be expected. Figure 6.7 shows the PDFs of fire sizes (E) for the two simulations of figure 6.6. At $p_g = 10^{-3}$ the distribution is approximately Gaussian, centered here around fire size 4800, but with a long, flat non-Gaussian tail extending to much smaller fires and a narrow, tall peak at very small fire sizes (off scale to the left in figure 6.7A). In this $p_f \ll 1$ regime, lowering the growth probability from $p_g = 10^{-3}$ to $p_g = 10^{-4}$ leads to a transition from a Gaussian distribution, with a relatively well-defined mean, to a power law of the form

$$f(E) = f_0 E^{-\alpha}, \quad \alpha > 0, \tag{6.1}$$

here with $\alpha = 1.07$. For such a power-law PDF it can be shown that the average fire size $\langle E \rangle$ is given by

$$\langle E \rangle = \frac{f_0}{2 - \alpha} \left[E_{\max}^{2-\alpha} - E_{\min}^{2-\alpha} \right], \tag{6.2}$$

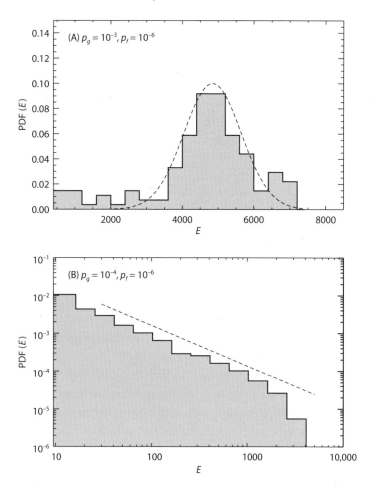

Figure 6.7. Probability distribution of fire size E for the two simulations of figure 6.6. The distribution in panel A is tolerably well fit by a Gaussian, except for its flat, low-amplitude tail extending to small fire sizes. The distribution in panel B is well described by a power law with index -1.07.

where E_{min} and E_{max} are the smallest and largest fires that can be produced by the simulation, here 1 and 10^4, respectively (see appendix B for the calculation of averages from a PDF). With $E_{min} \ll E_{max}$ (which is usually the case on large lattices) and

$\alpha < 2$, this is well approximated by

$$\langle E \rangle \simeq \frac{f_0 E_{max}^{2-\alpha}}{2 - \alpha}, \quad [\alpha < 2]. \tag{6.3}$$

This is because with $\alpha < 2$, the exponent $2 - \alpha$ in equation (6.2) is positive, so that the first term in the square brackets ends up much larger than the second. The opposite would be true if $\alpha > 2$. In the regime $p_g \ll 1$, $p_f \ll p_g$, the first case prevails, and therefore the largest fires dominate the evolution of the (eco)system.[4]

Whatever their shape, PDFs are defined such that $f(E)\,dE$ measures the occurrence probability of a fire of size between E and $E + dE$. In the ecologically realistic $p_g \ll 1$ regime, any one node contributes only one burned tree to a given fire; the situation was different in the sandpile model of the preceding chapter, where a node could topple repeatedly in the course of the same avalanche (see figure 5.5 if you're not convinced). Here, if a fire destroys E trees, it is because lightning hit somewhere within a cluster containing E connected trees. However, the probability that a cluster of size E is hit by randomly distributed lightning strikes is also proportional to the cluster size. Therefore, the PDF of cluster sizes must be distributed as $\propto E^{-(\alpha+1)}$ if the PDF of fire sizes is $\propto E^{-\alpha}$; with $\alpha = 1.07$ for the simulation of figure 6.7B, this implies that clusters of trees are distributed as a power law with index -2.07. Recall from chapter 4 that percolation clusters show a scale-invariant power-law size distribution only at the percolation threshold (see figure 4.7). Can we then conclude that the forest-fire lattice is at the percolation threshold?

It turns out to be significantly more complicated than that. Unlike in classical percolation, the tree density, equivalent to the occupation probability in percolation, is not constant across the lattice in the forest-fire model. This is illustrated in figure 6.8, showing a snapshot of the distribution of trees

---

[4]The same holds for the Earth's crust, with the largest earthquakes contributing the most to the relaxation of tectonic stresses; more on this in chapter 8.

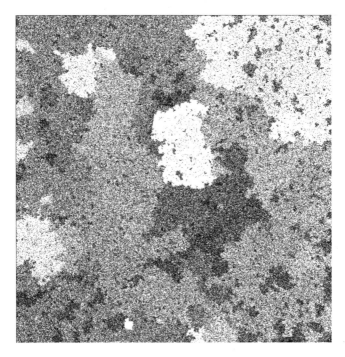

Figure 6.8. Snapshot of a forest-fire simulation on a 1024×1024 lattice, with parameters $p_g = 10^{-4}$ and $p_f = 10^{-7}$. Each small black dot is a tree, so that the resulting pointillistic gray shading provides a visual measure of tree density. Burning trees are plotted in red. Note how tree density, as measured visually by the level of gray shading, is approximately constant within contiguous domains, relatively well delineated but very irregularly shaped. The lighter areas are the scars of the more recent fires, and often contain dense clumps of surviving trees, corresponding to "holes" within the former clusters destroyed by fire. In this snapshot, two fires are burning, a large one near the lattice center and a smaller one near the bottom.

(black dots) in a $p_g = 10^{-4}$, $p_f = 10^{-7}$ simulation, now on a much larger 1024×1024 lattice. The mean density of trees is only approximately constant within irregularly shaped domains, with significant jumps occurring at the boundaries separating contiguous domains. These domains have been carved by prior fires having swept through the lattice. Tree growth, as mediated by Rule 1, is random but statistically homogeneous in space, so that the mean density of a given (large enough) domain is proportional to the time elapsed since the end of the last major fire having swept through that domain. Each individual domain behaves effectively as a separate percolation lattice, with slowly increasing occupation probability. Immediately following a fire, the occupation probability is close to zero, but grows linearly with time, eventually reaching the percolation threshold ($p_c = 0.4072$ for a Cartesian lattice with 8-neighbor connectivity). Recall that the likelihood of a single cluster taking over the lattice increases very rapidly once moving beyond this threshold (see figure 4.6), so that lightning, when and wherever it hits, is likely to wipe out the whole domain in a single fire. The shape and size of domains evolve slowly in the course of the simulation, because part of a domain may be destroyed by fire before reaching the percolation threshold (lightning hitting "early"), or by fusion with neighboring domains if both have exceeded significantly the percolation threshold prior to one igniting (lightning hitting "late"). Clearly, the PDF of fire sizes is determined by the past history of fires, going back at least a few p_g^{-1} iterations. Since typically $p_g \ll 1$, the system is said to exhibit long temporal correlations.

6.5 Back to Criticality

Running forest-fire model simulations for various combinations of growth and activation probabilities p_g and p_f, one soon realizes that in the portion of parameter space satisfying the double limit

$$p_f \ll p_g, \quad p_g \ll 1, \tag{6.4}$$

the PDF of fire sizes (and durations) always assumes a power-law shape. Moreover, in that regime the power-law index is always the same, and, for large enough lattices, is independent of lattice size. In other words, the corresponding values of α are universal, and involve no fine-tuning of control parameters. In this regime, the forest-fire model exhibits SOC.

In terms of the conditions for SOC behavior identified at the end of chapter 5, slow forcing is tree growth; the threshold instability (with respect to lightning strike) is the presence of a tree on the node hit by lightning; redistribution is the propagation of fire to neighboring trees. The system is open, because new trees are continuously added to the lattice, and dissipative, because a mechanism (fire) removes trees from the lattice.

But why should it matter whether wildfires represent an instance of SOC? It turns out to matter a lot when you decide to actively manage wildfires.

6.6 The Pros and Cons of Wildfire Management

As I write these lines, life is slowly returning to the 1,500,000+ acres of land (over 6000 square kilometers) charred by the spring 2016 Fort McMurray wildfire in Northern Alberta. It is currently lining up to rank as the costliest natural disaster in Canadian history. Amazingly enough, "only" two people died, in a car collision during the town's evacuation. Sometimes the toll gets worse. I lived in Colorado back in 1994 and vividly recall the Storm King Mountain wildfire near Glenwood Springs, which on July 6 claimed the lives of 14 firefighters who could not escape a rapidly moving fire front. And if this was not bad enough, a century ago an estimated 223 Northern Ontario residents suffered the same fate when half a dozen small communities were swept by the July 29, 1916 Matheson wildfire. The dangers of wildfires, and wildfire fighting, are not to be taken lightly. This is serious business.

In Canada, as in the United States, until recently and to some extent still now, wildfire management consisted in putting out potentially dangerous wildfires as quickly as possible, when the fire is still small. It sure seems to make a lot of sense. This type of fire management practice is easy to incorporate in the simulation code of figure 6.1. For example, introduce a time-dependent *extinction probability* (p_e) which decreases with the current number of burning trees (n_b) as

$$p_e(t) = \begin{cases} 0.2/n_b(t) & \text{if } n_b \leq 10, \\ 0 & \text{otherwise.} \end{cases} \tag{6.5}$$

Now when a fire is triggered and begins to grow, at every subsequent temporal iteration a probability test forces simultaneous extinction of all burning nodes with probability p_e. As the fire grows beyond 10 simultaneously burning nodes, this probability will become zero, reflecting the fact that real wildfires become very hard to extinguish once they really get going.

Considering that even large fires start off small, this procedure will clearly reduce the number of fires burning on the lattice over a set time span. Since the PDF of fire sizes has a power-law shape, however, most extinguished fires would have remained small anyway. Extinguishing them thus leaves more fuel for subsequent fires; whenever one manages to grow to a size where the probability of being extinguished goes to zero, as per equation (6.5), the forest is more densely packed with trees than it would have been had the earlier small fires not been extinguished. As a consequence, the total number of fires decreases, but the size of the largest fires may well increase! Not at all the desired outcome of good wildfire management. The Grand Challenge for this chapter leads you through a quantitative investigation of this phenomenon.

6.7 Exercises and Further Computational Explorations

1. The time series at the bottom of figure 6.4 shows very clear periodicity; can you determine what sets the period here?

2. Run two forest-fire simulations using the parameter values in figure 6.6. Make sure to run your simulations long enough to generate a few hundred fires at least. Calculate the fire measures E, P, and T, as in figure 6.3, and examine how these correlate against one another for your ensemble of fires. In both cases examine also whether fire size E correlates with the time elapsed since the end of the previous fire, or with the size of the previous fire.

3. The aim here is to have you test some modifications to the forest-fire model, and examine their impact. Work with a 100×100 lattice, and try at least one of the following (and the more the better!):

   a. Modify the Python source code of figure 6.1 so that it operates in "stop-and-go" rather than "running" mode, i.e., no tree is allowed to grow as long as a fire is burning anywhere on the lattice. In which parts of parameter space does this alter the global behavior of the model?

   b. Modify the Python source code of figure 6.1 so that fire propagates only to the 4 nearest neighbors top/down/right/left. Does this alter the global behavior of the model?

   c. Modify the Python source code of figure 6.1 so that the growth probability increases linearly with the number n of occupied neighboring nodes, for example $p_g \to p_g(1 + n)$. Does this alter the global behavior of the model?

   d. Modify the Python source code of figure 6.1 to introduce periodic boundary conditions horizontally and vertically (see appendix D for more detail on implementing such boundary conditions on a

lattice). Set $p_f = 10^{-5}$ and explore the types of
patterns generated at $p_g = 10^{-3}$ and 10^{-2}.

4. The forest-fire model is ideally suited to investigating
   an interesting variation on percolation, sometimes
   known as *dynamical percolation*. The idea is to replace
   the initial condition in the forest-fire model of
   figure 6.1 by a classical percolation lattice with
   occupation probability p (see the small Python code at
   the beginning of section 4.2). Now turn off tree growth
   and lightning, but as an initial condition, set on fire all
   nodes along the left edge of the lattice, and run the
   model until the fire extinguishes. Repeat the process for
   10 distinct random realizations of your percolation
   lattice, and keep track of the fraction of runs for which
   the fire reaches the right edge prior to extinction.
   Repeat for varying p, and construct a plot showing the
   fraction of "successful" realizations versus p. How does
   this plot compare to the bottom panel in figure 4.6?
   How would you estimate the percolation threshold p_c
   from such ensemble averaging?

5. The numerical implementation of the forest-fire model
   listed in figure 6.1 is extremely inefficient in many
   respects. For example, just consider the fact that every
   empty node on the lattice is subjected to the tree
   growth probability test at every temporal iteration; for
   an $N \times N$ lattice, since trees grow only on empty nodes
   (Rule 1), it would be much faster to "grow" a tree at
   $p_g \times N_e$ randomly selected empty nodes, where N_e is
   the number of empty nodes at the current iteration.
   Modify the Python source code of figure 6.1 to operate
   in this manner. And, if you feel up to some more
   serious coding, see exercise 5 in chapter 5 for
   more ideas.

6. And finally for the Grand Challenge: wildfire mitigation and management! The idea is to implement the strategy outlined in section 6.6 into the basic code of figure 6.1. Work off a 128×128 lattice in the SOC regime of figure 6.6: $p_g = 10^{-4}$ and $p_f = 10^{-6}$. Examine how the PDF of fire sizes varies as you increase the probability of extinction, i.e., replace the numerical factor 0.2 in equation (6.5) by the values 0.1, 0.2, 0.3, and 0.5. Run the simulations for the same number of temporal iterations in all cases. Is the PDF getting steeper or flatter as the probability of extinction increases? How about the size of the largest fires? How would you go about designing an "optimal" wildfire management strategy in the context of this model? Note that in the context of this Grand Challenge, you will be trying to obtain accurate determinations of the power-law index of the PDFs at each extinction probability; you may consider calculating this index following the maximum likelihood approach described in appendix B.6. Make sure to exclude the initial transient phase from your analyses, and to push the simulations far enough in time to have many hundreds of fires to build your PDFs from, even in the simulation with the fewest fires.

6.8 Further Reading

The forest-fire model introduced in this chapter is due to

Drossel, B., and Schwabl, F., "Self-organized critical forest-fire model," *Phys. Rev. Lett.*, **69**(11), 1629–1632 (1992).

A comprehensive review of its properties can be found in

Hergarten, S., "Wildfires and the forest-fire model," in *Self-organized Criticality Systems*, ed. M.J. Aschwanden, Berlin: Open Academic Press, 357–378 (2013).

I know of no good textbook dedicated to the mathematical modeling of wildfires, but the topic is sometimes covered in textbooks on mathematical modeling in general. I did find the Wikipedia page on wildfire modeling well balanced and quite informative, and it also includes many good references to the technical literature (consulted November 2014):

http://en.wikipedia.org/wiki/Wildfire_modeling.

On the comparison of real wildfire data with a suitably tailored SOC-type model akin to that considered in this chapter, including fire management strategies, see

Yoder, M.R., Turcotte, D.L., and Rindle, J.B., "Forest-fire model with natural fire resistance," *Phys. Rev. E*, **83**, 046118 (2011).

7

TRAFFIC JAMS

Avalanches on a sandpile and forest fires on a lattice both represent a form of complex collective behavior emerging from simple interactions between a large number of equally simple interacting elements. There is no directed purpose in the toppling of a sand grain, or the ignition of a tree by a neighboring burning tree.

Complex collective behavior can also emerge from the interactions of system elements that do behave in a purposeful manner, and in some cases this collective behavior may even appear to run counter to the purpose driving these individual interacting elements. The occurrence of traffic jams in the flow of moving automobiles is a fascinating example, and is the focus of this chapter.

7.1 Model Definition

The basic model design is once again conceptually quite simple. A line of N cars is moving in the same direction along a single-lane one-way road. The agents driving the cars slow down if they come too close to the car ahead of them, accelerate if the distance allows it, and respect the speed limit. No passing or reversing is allowed. Think about it a bit; these are pretty realistic and conventional "driving rules." More specifically, and with the positions and speed of the kth car at time t_n henceforth denoted by x_k^n and v_k^n ($k = 0, \ldots, N - 1$), the speed adjustment rules are the following:

1. At each time step (n), each driver (k) "calculates" (or eyeballs) the distance δ to the car ahead:

$$\delta = x_{k+1}^n - x_k^n. \tag{7.1}$$

2. If $\delta < 5$, the car slows down:

$$v_k^{n+1} = v_k^n - 3. \tag{7.2}$$

3. If $\delta > 5$, the car speeds up:

$$v_k^{n+1} = v_k^n + 1. \tag{7.3}$$

4. The car speed must always remain bound in $[0, 10]$, 10 being the speed limit, and the lower bound precluding reversing.

5. Each car moves according to the standard prescription for uniform speed (i.e., uniform within a given temporal iteration):

$$x_k^{n+1} = x_k^n + v_k^n \times \Delta t. \tag{7.4}$$

In all that follows, we set $\Delta t = 1$ without any loss of generality.

6. And here is the crux. Every once in a while, due to an incoming text message, a change of CD, a squirrel crossing the road, or just for the sheer fun of being a royal pain in the lower backside, some random bozo agent ($k = r$) slams on the brakes:

$$v_r^{n+1} = v_r^n - 3, \quad r \in [0, N - 1]. \tag{7.5}$$

```
1  # DISCRETE TRAFFIC MODEL ON A ONE-WAY STRAIGHT ROAD
2  import numpy as np
3  import matplotlib.pyplot as plt
4  #-----------------------------------------------------------------------
5  N=300                             # number of cars
6  p_bozo=0.1                        # probability of random braking
7  n_iter=2000                       # number of temporal iterations
8  #-----------------------------------------------------------------------
9  v=np.zeros(N)                     # zero initial speeds for all cars
10 x=np.zeros(N)                     # car positions
11 mean_v=np.zeros(n_iter)           # time series of mean speed
12 x[0]=1                            # first car at x=1
13 for k in range(1,N):              # initialize car positions
14     x[k]=x[k-1]+np.floor(np.random.uniform(3.,14.))
15
16 for iterate in range(0,n_iter):          # temporal loop
17
18     for k in range (0,N-1):              # first car loop: update speeds
19         dx=x[k+1]-x[k]                   # distance to next car ahead
20         if dx < 5:                       # too close: slow down
21             v[k]=max(0,v[k]-3)
22         if dx > 5:                       # far enough: speed up
23             v[k]=min(10,v[k]+1)
24         if x[N-1]-x[N-2] <= 10:          # special case: lead car
25             v[N-1]=min(10,v[N-1]+1)
26
27     for k in range(0,N):                 # second car loop: braking
28         if np.random.random() <= p_bozo: # some bozo slams the brakes
29             v[k]=max(0,v[k]-3)
30
31     for k in range(0,N-1):               # third car loop: update positions
32         x[k]=min(x[k]+v[k],x[k+1]-1)     # with no-crash safety test
33     x[N-1]+=v[N-1]                       # special case: lead car
34
35     mean_v[iterate]=(v.sum())/N
36     print("iteration {0}, mean speed {1}.".format(iter,mean_v[iterate]))
37 # end of temporal loop
38 plt.plot(range(0,n_iter),mean_v)         # plot mean speed vs iteration
39 plt.axis([0,n_iter,0.,10.])
40 plt.xlabel('Iteration')
41 plt.ylabel('Mean car speed')
42 plt.show()
43 # END
```

Figure 7.1. Python source code for the simple automobile traffic model introduced in this chapter.

Unlike with normal braking, here this (hopefully) rare, random occurrence takes place *independently* of the distance to the car ahead. This is also the only one of the six driving rules which is not fully deterministic.

7.2 Numerical Implementation

The Python source code listed in figure 7.1 offers a simple implementation of the above traffic model. Take note of the following:

1. The simulation is once again structured around an outer temporal loop (starting at line 16), enclosing three sequential inner loops over the N cars (starting at lines 18, 27, 31).
2. Car positions are initialized as random-valued positive *increments*, here in the range $3 \leq x_{k+1} - x_k \leq 17$ (line 14), for a mean intercar distance of 10 units. This procedure ensures that $x_1 < x_2 < x_3 < \cdots < x_N$.
3. First, the changes in car velocities are computed for all cars in the first two inner loops, and only then are the car positions calculated and the array x updated, in the third inner loop (lines 31–32).
4. Safety tests using the Python functions min and max ensure that the speed cannot exceed 10 (lines 23 and 25), or fall below 0 (line 21).
5. Similarly, a "safety test" (line 32) ensures that no car can get closer than 1 unit to the car ahead.
6. Car number N, in the lead, does not have a car ahead of itself; consequently it adjusts its speed according to the distance to the *following* car (line 24–25).

7.3 A Representative Simulation

Figures 7.2 and 7.3 show results for a typical simulation, here for an ensemble of 300 cars initially at rest and distributed randomly, with a mean spacing of 10 units. This is actually the same initial condition set up in the source code of figure 7.1. Both figures show the trajectories, position versus time, for all cars (figure 7.2) or a subset thereof (figure 7.3). The first figure focuses on the

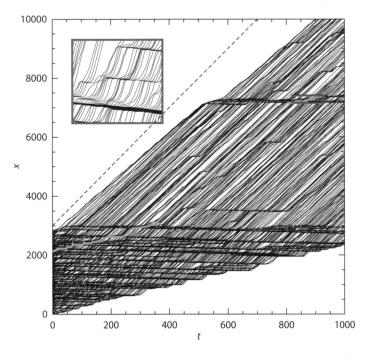

Figure 7.2. Trajectories of all cars in the simulation, defined as the variation of their position (vertical axis) versus time (horizontal axis), as produced by the Python code of figure 7.1, with all 300 cars initially at rest. The simulation evolves according to two fairly distinct phases, the first being one of ubiquitous traffic jams, transiting toward a state in which all cars move at the same average speed, but with traffic jams of varying sizes still occurring intermittently. The green line shows the trajectory of the car initially located a quarter of the way behind the leading car. The dotted line shows the slope corresponding to the speed limit $v = 10$. The inset zooms in on a large traffic jam, and shows than even in a jam, car trajectories never cross (no passing allowed on a single-lane one-way road!).

first 1000 temporal iterations of the simulation, while the second extends much farther, to 10^4 iterations. On such plots, horizontal streaks are symptomatic of cars at rest, i.e., traffic jams.

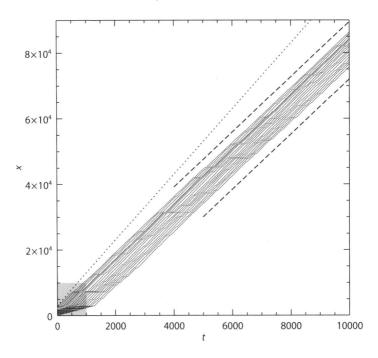

Figure 7.3. Same as figure 7.2, but covering a temporal interval 10 times longer, and with only 11 car trajectories plotted, for clarity. The gray-shaded area in the lower left is the range covered by figure 7.2. The dotted line is again the trajectory of a car moving at the maximum speed $v = 10$, and the two dashed lines mark the average speed of the ensemble of cars in the fluid phase of the simulation.

Early in the simulation (figure 7.2), traffic is a total mess because the initial spacing between cars is too small. Cars are continuously braking, triggering more braking in the cars following. As the first cars ahead of the line start to increase their speed and move ahead of the mess, cars behind them eventually do the same, until all cars have managed to pick up speed and increase the distance between each other, which occurs here after about 1300 temporal iterations. Loosely speaking, we can define this as the beginning of the "fluid" phase of the simulation, whereas the hopeless jam characterizing the first $\sim 10^3$ iterations will be referred to as the "solid" phase. It is clear on both these figures that even when the simulation is far into its fluid phase, jams of varying sizes still occur intermittently. Most of these are caused by a random bozo braking, but such an individual perturbation will sometimes have little effect, while at other times a jam involving almost all cars is produced, for example the jam beginning at $(x, t) \simeq (7000, 500)$ in figure 7.2. Note that here, with 300 cars and a bozo probability of 0.1, on average 30 random braking events take place at every temporal iteration, which is substantial.

The inset in figure 7.2 shows the trajectory of a specific car, in green, having just managed to free itself from a major jam having affected nearly the whole system, and subsequently hitting the back of, and later extracting itself from, two smaller jams. Upon careful examination of this inset, it becomes clear that individual cars are either moving at or close to the speed limit, or are at rest or nearly so, stuck in a jam. It is also noteworthy that the temporal duration of a jam is substantially longer than the time any single car spends stuck in it (take another look at the green trajectory in the inset to figure 7.2). This happens because cars free themselves from the jam one by one at its downstream end, while other cars pile up at its upstream end. As a consequence, once triggered the jam grows backward in x with time, even though no car ever moves backward here.

In position versus time plots such as in figures 7.2 and 7.3, the slope of the car trajectories gives the average speed of the ensemble of cars. This is indicated by the two parallel dashed lines bracketing the car trajectories in figure 7.3. The corresponding slope is very well defined and remains constant in the fluid phase of the simulation. Note however that it is significantly *smaller* than the slope expected for a car moving uniformly at the speed limit, which is indicated here by the dotted line. In other words, even though cars could all, in principle, move at the speed limit, through their interactions they settle in a mean state where their ensemble average speed is significantly smaller than the speed limit.[1] You should recognize this type of collective "suboptimality" as something we have encountered already, and if not, go take a look again at figure 5.3.

7.4 Model Behavior

We need to get a bit more quantitative in our attempts to understand how this model behaves. Two interesting global quantities are the *mean speed* for all cars,

$$\langle v \rangle = \frac{1}{N} \sum_{k=0}^{N-1} v_k^n, \qquad (7.6)$$

and the *mean distance* between successive cars in the lineup,

$$\langle \delta \rangle = \frac{1}{N-1} \sum_{k=0}^{N-2} (x_{k+1}^n - x_k^n) = \frac{x_{N-1}^n - x_0^n}{N-1}. \qquad (7.7)$$

The *mean density* of cars is simply the inverse ratio of this expression:

$$\rho = \frac{N-1}{x_{N-1}^n - x_0^n}. \qquad (7.8)$$

Knowing these two quantities, one can compute the *car flux* (Φ):

$$\Phi = \rho \times \langle v \rangle. \qquad (7.9)$$

This measures the average number of cars passing a given position x^* per unit time.

Figure 7.4 shows time series of the three quantities $\langle v \rangle$, ρ, and Φ for the simulation of figure 7.2. All three vary markedly in the early part of the simulation, until the transition to the fluid

---

[1]This is *not* due to random braking of individual cars; with a bozo probability of 0.1 and reacceleration to full speed requiring three iterations, the average speed of an isolated (noninteracting) car would be 9.4 here.

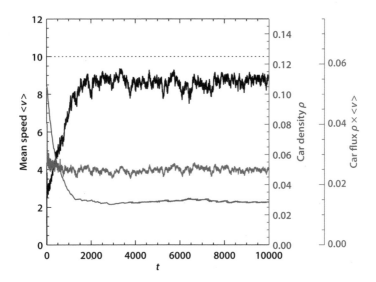

Figure 7.4. Time series of mean speed $\langle v \rangle$, mean density ρ, and car flux Φ in the simulation of figures 7.2 and 7.6. Even though the fluid phase begins around $t = 1300$, statistical stationarity is reached much later, around $t = 2000$.

phase at $t \simeq 1300$. Note, however, that the mean density and flux of cars only really stabilize starting around $t \simeq 2000$. This indicates that reaching a statistically stationary state still requires a significant amount of time after transiting from the solid to the fluid phase.

A noteworthy property of this statistically stationary state is that its global characteristics, such as mean speed and density, are independent of the initial condition for the simulation. The mess of monster traffic jams characterizing what we dubbed the solid phase of the simulation certainly suggests that the initial condition imposed here is far from optimal, in terms of getting the traffic going. Nonetheless, cook up whichever initial condition you can think of, with the traffic rules used here, and for a large enough number of cars, the system always stabilizes

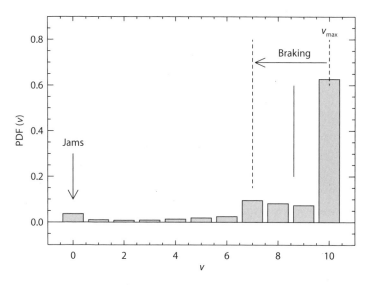

Figure 7.5. Probability density function of car speeds, built from the speeds of all cars at each temporal iteration far into the fluid phase ($t > 3500$) of the simulation plotted in figures 7.2 and 7.3. The vertical line segment at $v \simeq 8.6$ indicates the mean speed, and the secondary peak at $v = 7$ is a direct consequence of the braking rule ($v \rightarrow v - 3$) for cars moving at the speed limit, which dominate the distribution (see text).

at the same statistically stationary values of $\langle v \rangle$, ρ, and Φ as in figure 7.4.

The evolution toward such robust mean car speeds (and densities) would also suggest that most cars end up traveling most of the time at or near that speed; in other words the distribution of car speeds is Gaussian-like and centered on its mean value $\langle v \rangle$. This is not at all the case, as one can immediately see from figure 7.5. This shows the PDF of car speeds,[2] built from all cars at all iterations far into the statistically stationary fluid phase

---

[2]Since car speed is defined as an integer in the range $0 \leq v_k^n \leq 10$, this distribution is fundamentally restricted to 11 bins.

of the simulation ($t > 3500$). The distribution is in no way Gaussian, or even symmetrical about its mean value (vertical line segment at $\simeq 8.6$), but instead spans the whole allowed range, with its peak at $v = 10$ and secondary peaks at $v = 7$ and $v = 0$. The $v = 7$ peak is a direct consequence of the braking rule (equation (7.2)) (which decrements speed by 3 units) acting on the primary peak at $v = 10$.

What this distribution expresses is worth detailing and reflecting upon. Cars spend over 60% of their time moving at the speed limit $v = 10$, and only 4% of their time stuck in a traffic jam of whatever size, which is really not so bad after all (although for most people, myself included, the stress level generated by the time spent in the jams would be disproportionately much higher). Note also that while a "mean car speed" can be defined unambiguously from a mathematical point of view, in itself it does not provide very useful information regarding the state of a specific car, even in a statistical sense. This stands in contrast to a situation where the car speeds are distributed as a Gaussian, in which case the mean speed also coincides with the most probable speed. This is not the case in figure 7.5, where the most probable speed is $v = 10$, significantly higher than the mean speed.

7.5 Traffic Jams as Avalanches

You probably have already figured out that the buildup of a traffic jam in these simulations is akin to an avalanche of successive braking events. Moreover, at the dynamical level, nothing fundamentally distinguishes small jams from large ones; all that changes is the number of cars involved. Could we not then expect jams to exhibit some form of scale invariance? Let's look into that.

Some care is warranted in defining the "size" of a traffic jam; the number of cars involved is obviously an important factor, but so is the temporal duration of the jam, which, as we have already noted, is typically larger than the time any individual car spends stuck in it. A jam is a pseudo-object, in that cars are continuously piling up at the back of the jam, and others are removed at its front. Much like a waterfall, which retains its shape despite the fact that water is flowing through it, a large traffic jam retains its "identity" for a length of time usually much larger than the time any one car spends moving through it. Traffic jams are spatiotemporal structures and must be treated as such.

Consider the following procedure: We build a rectangular pixelated "image" where there are as many pixels horizontally as there are cars, and as many pixels vertically as there are temporal iterations. Each pixel (k, n) in the 2-D image is assigned an integer value between 0 and 10, set equal to the speed of car k at iteration n, i.e., v_k^n. Figure 7.6 shows the results of this procedure, in the form of three successive 1000-iteration-long blocks laid side by side, with color encoding the speed, according to the scale on the right. These representations illustrate well the fact that traffic jams are structures that exist in space and time, and their backward propagation, one car at a time, becomes particularly striking.

Now, the idea is to define a traffic jam as a cluster of pixels with value zero, contiguous in car number space (horizontally) and time (vertically). Figure 7.7 illustrates the idea, for a 300-iteration-long segment corresponding to the middle portion of the central column in figure 7.6. Clusters of halted cars evidently span a wide range of sizes, going from a single pixel up to slanted structures stretching over many hundreds of iterations and collecting in excess of 10^3 pixels. In some cases, pixels that appear to "belong" to the same jam, as per their location along the same slanted streak of pixels, end up broken into a string of smaller groups. Some smaller jams also occasionally merge into larger ones, but the model's governing rules make it difficult for a jam to spawn secondary branches, a rare occurrence restricted to very small jams. Figure 7.7 seriously begins to smell of scale invariance. And you will undoubtedly recall that, back in chapter 4 when

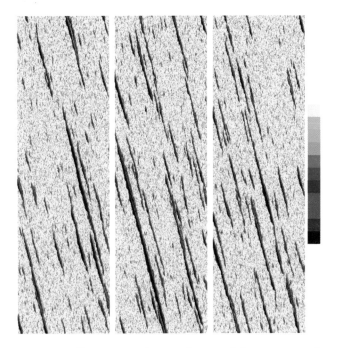

Figure 7.6. Traffic jams in the simulation of figures 7.2 and 7.3. What is plotted on each of the three color-coded images is the speed of the cars as a function of car number (running horizontally) and time (running vertically from bottom to top), for three successive 1000-iteration chunks of the simulation in its fluid phase. Zero speed is black, going through blue and red up to $v = 10$ in yellow, as per the 11-step color scale on the right.

investigating percolation, we introduced an algorithm (based in fact on the forest-fire model of chapter 6), that can assign a unique numerical tag to each such cluster. Because of the tendency of jams to shift backward one car per time step, here we define contact with any of a pixel's 8 nearest neighbors, i.e., including diagonally, as the criterion for tagging pixels to the same clusters. This involves only a minor modification to the cluster-tagging code of figure 4.3.

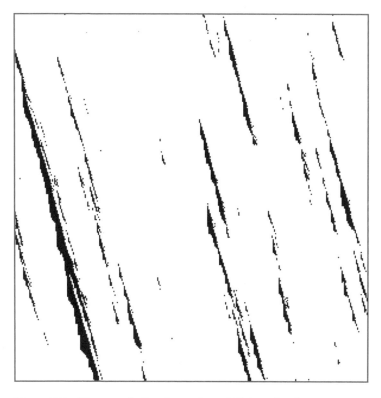

Figure 7.7. Clusters of $v_k^n = 0$ cars for a 300-iteration-long segment of the simulation corresponding to the middle part of the central column in figure 7.6. Note the merging of small jams into the larger jam running on the left.

Figure 7.8 shows the PDF of traffic jam sizes, built from the 3441 distinct jams tagged in the last 8000 iterations of our now familiar simulation of figures 7.2 and 7.3. The first 2000 iterations have been omitted so as to restrict the statistics to the stationary fluid phase. Once again, the sizes are distributed as a well-defined power law, spanning here over two orders of magnitude in size, with logarithmic slope -1.58. This power-law form supports—but does not rigorously prove—our growing suspicion that traffic jams are scale-invariant spatiotemporal structures.

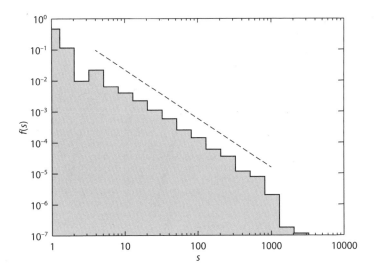

Figure 7.8. Probability density function for the sizes of the 3441 distinct traffic jams tagged in the last 8000 iterations of the simulation plotted in figures 7.2 and 7.3. The distribution is well fit by a power law with index −1.58.

7.6 Car Traffic as a SOC System?

Scale invariance is a hallmark of critical systems, but its presence is certainly not *proof* of the presence of criticality; the aggregates of chapter 3 were scale invariant, but the DLA process has nothing to do with criticality.[3] On the other hand, our traffic model does show a key defining feature of critical systems: in its statistically stationary state, one small perturbation (a randomly braking bozo) has a finite probability of affecting the whole system, through the triggering of a jam bringing all cars to a grinding halt, from first to last. This is akin to a lattice at the percolation threshold, where the appearance of a single additional occupied

---

[3]Or does it? If you are keen on the issue, read and reflect upon the Witten and Sanders paper cited at the end of chapter 3.

node can produce a cluster spanning the whole lattice. Moreover, and now unlike percolation, here this state arises autonomously through the interactions between a large number of moving cars. If it *is* criticality, then it is also SOC.

The lofty objective of traffic engineering is to ensure a smooth flow of automotive traffic, subject to the additional desirable, practical goal that all participating drivers get to where they want to go as quickly and painlessly as possible.[4] One would strongly suspect that traffic jams represent a major obstacle toward this goal. Can these traffic jams be avoided? Obviously, one possibility is to ensure a spacing between cars that is large enough for a random braker to have time to accelerate back to the speed limit before the next car behind has caught up and is forced to brake. However, such a state would be characterized by a low density of cars, and therefore a low flux even if all cars fly along smoothly at the speed limit. If the objective is to get a very large volume of commuter traffic into town, this will not do. One could try the opposite approach and pack cars as closely as possible behind one another, thus reaching high densities and therefore high flux; but such a state will always produce a huge jam as soon as a bozo decides to brake for nothing, causing a massive slowdown of a great many cars, with the flux dropping precipitously as a consequence, and recovery to a fluid phase is a lengthy process. Is there a working solution to this flux maximization problem? The answer is thought to be yes, and we have been staring at it all along.

It has been conjectured that the stationary state attained by these traffic simulations, despite the jams of all sizes occurring across the system, actually *maximizes* the flux of cars in the presence of random brake-slamming bozos, as compared to any

---

[4]My home town, Montréal, seems to operate under a different method; or perhaps there is just no method at all.

other carefully engineered traffic state.[5] In other words, a scale-invariant distribution of traffic jams is the system's emergent strategy for minimizing the *global* impact of randomly braking bozos. Certainly nothing of this sort could have been anticipated on the basis of the simple traffic rules defining the model. You actually get to test some aspects of this remarkable conjecture in some of the computational explorations suggested below.

7.7 Exercises and Further Computational Explorations

1. It was stated back in chapter 5 that the necessary conditions for SOC were (in short) a slowly driven open system subjected to a self-stabilizing local threshold instability. Can you identify these elements in the traffic model considered in this chapter? How could you argue that this is yet another instance of an open dissipative system?

2. This one lets you explore some parameter dependencies of the traffic model introduced in this chapter.

   a. Generate a series of traffic simulations with varying numbers of cars (30, 100, 300, 1000, and 3000, say). Investigate whether the mean speed, density, and car flux in the fluid phase depend on the total number of cars.

   b. Use the code of figure 7.1 to produce a set of traffic simulations with an increased probability of random braking (variable p_bozo), but otherwise identical. Examine the effect on the mean speed attained in the fluid phase of the simulation. Do you always see a reasonably well-defined transition from "solid" to "fluid"?

---

[5] This is a conjecture, in the sense that no one has yet been able to rigorously prove it, as far as I know anyway; but no one has managed to offer a clear counterexample either.

3. Try to engineer an initial condition which will minimize the duration of the "solid" phase of traffic. The idea is of course to distribute a set number of cars on a set length of road; what you can play with are the positions and initial speeds of the cars. Are the mean speed and car density that are attained in the statistically stationary fluid state dependent on the initial conditions?

4. Change the acceleration and braking rules (i.e., the magnitude of the increment and decrement in speed), and examine the impact of such changes on the upstream/downstream motion of jams. Can you infer a simple mathematical relationship between these model parameters (microscopic rules) and the motion of jams (macroscopic behavior)?

5. A commuter's nightmare version of our traffic jam model can be produced by having the cars move along a circular one-way ring road. Your first task is to modify the Python code of figure 7.1 accordingly. Think this one through carefully; you can do this by changing a single line of code in figure 7.1, once you define the length of the road perimeter. How does the model behave as compared to the original straight-road version introduced in this chapter?

6. The Grand Challenge for this chapter is two pronged. You have to work with the ring-road version of the model, as described in the preceding exercise.

   a. Examine how the mean speed and car flux in the statistically stationary state vary as functions of car density (as controlled by the number of cars placed on the ring road) for a fixed road perimeter. Does this remind you of something? If not, go back and reread chapter 4, then come back and determine the percolation threshold for this ring road.

b. The dynamical rules defining the traffic model introduced in this chapter are invariant under an inversion of car velocities, $v_k \to -v_k$ for all k. Modify the ring-road version of the model so that initial car speeds are set randomly at either $+1$ or -1 equiprobably. Adjust the driving rules accordingly, and add a "chicken" rule: whenever two cars are about to collide face on, reverse the speed of the slowest car (and pick a direction randomly if they have the same speed). Use an initial car density that is sufficiently low for a fluid phase to be eventually attained (as per your investigations in (a)). Carry out an ensemble of simulations with distinct random initializations, and verify that in the end state, both senses of driving (clockwise and counterclockwise) are equally probable. This represents an instance of *symmetry breaking*: nothing in the dynamical rules favors one sense of rotation over the other; the direction of the global flow of cars emerges from the (symmetrical) dynamical rules acting on the low amplitude "noise" of the initial condition.

7.8 Further Reading

There exists a vast literature on the mathematical modeling of traffic flow. The following (advanced) textbook offers a good survey of the current state-of-the-art:

Treiber, M., and Kesting, A., *Traffic Flow Dynamics*, Springer (2013).

The traffic model studied in this chapter essentially follows that proposed by

Nagel, K., and Paczuski, M., "Emergent traffic jams," *Phys. Rev. E*, **51**, 2909 (1995);

but see also chapter 3 in

Resnick, M., *Turtles, Termites, and Traffic Jams*, MIT Press (1994).

My first encounter with the mathematical modeling of traffic jams was in chapter 5 of the following delightful book, which the mathematically inclined should not miss:

Beltrami, E., *Mathematics for Dynamical Modeling*, San Diego: Academic Press (1987).

8

EARTHQUAKES

Earthquakes are *scary*, because they are powerful and (as yet) unpredictable, and can have consequences going far beyond rattling the ground under our feet; just in recent years, think of the earthquake-triggered December 2004 killer tsunami in the Indian ocean, or the March 2011 failure of the Fukushima nuclear power plant in Japan, or the hundreds of thousands of people left homeless by the April 2015 earthquake in Nepal. This is serious business.

It is now understood that Earth's crust is broken into a dozen or so major tectonic plates, about 100 km thick, floating on a deep very viscous fluid layer of molten rocks called the mantle. Horizontal fluid motions are ubiquitous in the outer mantle, due to thermally driven convection in Earth's interior. These flows produce a horizontally oriented viscous force at the bottom of tectonic plates, which is opposed by static friction at the boundaries between adjacent plates moving relative to one another. These regions of high static stress are known as *fault lines*. As the viscous force builds up, the rock first deforms elastically, but there comes a point where static friction and deformation can no longer offset forcing. The plates abruptly move, producing what we call an earthquake.

The energy released by earthquakes is quantified by their magnitude m, essentially a logarithmic measure of seismic wave amplitudes. A long-known, remarkable property of earthquake energy releases is that the distribution of their magnitudes takes the form of a power law. More specifically, the number N of earthquakes having a magnitude larger than m in a given area and time interval is given by the celebrated Gutenberg–Richter law:

$$N(> m) \propto m^{-b}, \qquad (8.1)$$

where $b \simeq 1$ in most locations.[1] This power law is taken to reflect scale invariance in the dynamics of earthquakes, a property that can be reproduced using a simple mechanical model, to which we now turn.

8.1 The Burridge–Knopoff Model

The Burridge–Knopoff stick–slip model of seismic faults is a mechanical construct defined as a 2-D array of blocks interconnected by springs to their 4 nearest neighbors, sandwiched in the vertical between two flat plates (see figure 8.1). Each block can be tagged by a pair of indices (i, j) measuring its relative position in x and y in the array. The blocks rest on the bottom plate and are each connected to the top plate by another set of leaf springs. Figure 8.1 illustrates this arrangement for a block (i, j) and its 4 nearest neighbors $(i - 1, j)$, $(i + 1, j)$, $(i, j - 1)$, and $(i, j + 1)$. The bottom plate is assumed to be at rest, but the top plate moves in the positive x-direction at a constant speed V. This is the model's analogue to the moving mantle fluid and the viscous force it impresses on the plates. The motion of the upper plate will gradually stretch the leaf springs, thus inexorably increasing the x-component of the force acting on each block.

---

[1] Equation 8.1 is a *cumulative* PDF; the usual bin-count-based PDF would be $\propto m^{-(b+1)}$. If needed, see appendix B for more on cumulative PDFs.

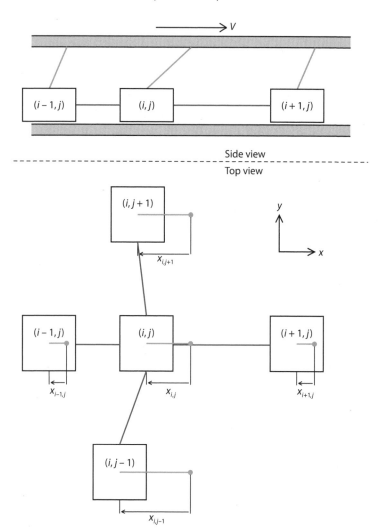

Figure 8.1. The Burridge–Knopoff sliding-block model of earth-quakes, displayed here in top and side views. The bottom plate is assumed fixed and the top plate moves with a constant speed V. Leaf springs are traced in green, and interblock springs in red. The block displacements (x) are measured from the anchoring points of the leaf spring on the top moving plate, indicated by green dots in the top-view diagram.

The model assumes that Hooke's law holds, meaning that the force is linearly proportional to the stretching of each spring:

$$F_x = K \Delta x, \tag{8.2}$$

where K is the spring constant, and the displacement $\Delta x \equiv x_{i,j}$ is here set equal to the distance between the block center and the anchoring point of its leaf spring on the top plate (see figure 8.1). The spring constants of the interblock springs and leaf springs are not necessarily the same, and are respectively denoted K and K_L in what follows.

The x-component of the total force acting on block (i, j) is given by the sum of the contributions from the spring connected to the 4 nearest neighbors, plus that of the leaf spring:

$$\begin{aligned}
F_{i,j}^n &= K(x_{i-1,j}^n - x_{i,j}^n) + K(x_{i+1,j}^n - x_{i,j}^n) + K(x_{i,j-1}^n - x_{i,j}^n) \\
&\quad + K(x_{i,j+1}^n - x_{i,j}^n) - K_L x_{i,j}^n \\
&= K(x_{i-1,j}^n + x_{i+1,j}^n + x_{i,j-1}^n + x_{i,j+1}^n - 4x_{i,j}^n) - K_L x_{i,j}^n,
\end{aligned} \tag{8.3}$$

where, in anticipation of developments to follow, the superscript n indicates time.[2] Missing from this expression is the static friction force acting between the block and the lower plate on which it rests. As long as this can equilibrate the force mediated by the springs, given by equation (8.3), every block in the system remains at rest.

Because the displacement of the top plate increases inexorably the force transmitted by the leaf springs to the blocks, there will inevitably come a point when the friction force cannot counteract the spring forces, and a block will slip. The key idea here is that

---

[2] Because the displacements x are measured from the anchoring points of the leaf spring, in general they will be negative quantities (like the five illustrative displacements in figure 8.1). Consequently, a term like $-K_L x_{i,j}^n$ in equation (8.3) is positive signed, indicating that the leaf spring pulls the block in the positive x-direction, as it should.

upon slippage, the block (i, j) rapidly settles at an equilibrium position where the net spring force is zero:

$$F_{i,j}^{n+1} = K(x_{i-1,j}^n + x_{i+1,j}^n + x_{i,j-1}^n + x_{i,j+1}^n - 4x_{i,j}^{n+1})$$

$$-K_L x_{i,j}^{n+1} = 0, \tag{8.4}$$

again with superscript $n + 1$ denoting the time after slippage. The change in the total spring force acting on block (i, j) is thus

$$\delta F_{i,j} \equiv F_{i,j}^{n+1} - F_{i,j}^n = (4K + K_L)(x_{i,j}^{n+1} - x_{i,j}^n). \tag{8.5}$$

Since $F_{i,j}^{n+1} = 0$ by prior assumption (namely, equation (8.4)), the right-hand side of this expression must be equal to $-F_{i,j}^n$. Consider now the neighboring block $(i + 1, j)$, say. Assuming that only block (i, j) has undergone slippage, the corresponding change in the total force acting on block $(i + 1, j)$ is simply

$$\delta F_{i+1,j} = K(x_{i,j}^{n+1} - x_{i,j}^n). \tag{8.6}$$

This must be equal to $\delta F_{i,j}$, as per Sir Isaac Newton's celebrated action–reaction dynamical law; equation (8.5) can thus be used to substitute for $x_{i,j}^{n+1} - x_{i,j}^n$ in equation (8.6), which immediately leads to

$$\delta F_{i+1,j} = \alpha F_{i,j}^n, \tag{8.7}$$

where

$$\alpha = \frac{K}{4K + K_L}. \tag{8.8}$$

Therefore, the force on block $(i + 1, j)$ varies by an amount proportional to the force acting on block (i, j) before slippage, a result which also holds for the other three nearest-neighbor blocks.[3] The numerical value of the proportionality constant α

---

[3] In general, the slipping block would also move in the y-direction, unless $x_{i,j+1} = x_{i,j-1}$; this can be ignored here because the y-displacement will average

is set by the ratio of spring constants; note in particular that

$$\lim_{K \ll K_L} \alpha = 0, \qquad \lim_{K \gg K_L} \alpha = \frac{1}{4}. \qquad (8.9)$$

It turns out to be possible to design a simple sandpile-like model in which the rules can be unambiguously related to the physical laws at play in the Burridge–Knopoff sliding block model. The key is to use the total force $F_{i,j}$ acting on block (i, j) as a nodal variable, rather than its position.

As with the 1-D sandpile model considered in chapter 5, the Olami–Feder–Christensen (hereafter OFC) model is a lattice-based CA-like system evolving according to simple rules discrete in space and time. In keeping with the Burridge–Knopoff sliding block picture, we consider here a 2-D Cartesian lattice made up of $N \times N$ nodes with top+down+right+left neighbor connectivity. This lattice is used to discretize a real-valued variable $F_{i,j}^n$, where the subscript pair (i, j) identifies each node and the superscript n now denotes a discrete temporal iteration.

The nodal variable is subjected to a deterministic forcing mechanism, whereby at each temporal iteration, a small increment δF is added to the force variable F at every node on the lattice:

$$F_{i,j}^{n+1} = F_{i,j}^n + \delta F \quad \text{for all } i, j. \qquad (8.10)$$

This captures the slow displacement of the top plate in the Burridge–Knopoff model, which inexorably increases the force transmitted to all blocks through their leaf spring. Whenever the total force on the block exceeds some preset threshold F_c,

$$F_{i,j}^n > F_c, \qquad (8.11)$$

---

to zero after many slipping events, a consequence of the fact that the forcing by the upper plate is aligned with the x-direction. Also, note that equation (8.8) is valid only for "interior" blocks; those at the edges and corners of the block system would have $\alpha = K/(3K + K_L)$ and $\alpha = K/(2K + K_L)$, respectively.

corresponding physically to the friction force between the blocks
and the bottom plate, the node relaxes to a zero-force state by
redistribution to its nearest neighbors:

$$F_{i,j}^{n+1} = 0, \tag{8.12}$$

$$F_{nn}^{n+1} = F_{nn}^{n} + \alpha F_{i,j}^{n}, \quad 0 \leq \alpha \leq 0.25, \tag{8.13}$$

where $nn \equiv [(i+1, j), (i-1, j), (i, j+1), (i, j-1)]$, and
α is in fact the very same proportionality constant appearing
in equations (8.7) and (8.8), i.e., it measures the fraction of
the force acting on the unstable node that is lost to the upper
plate, rather than being redistributed to the nearest neighbors.
This redistribution evidently restores local stability to node (i, j),
but as in the sandpile model of chapter 5, one or more of the
nearest neighbors can be pushed beyond the stability threshold
by the redistribution of the nodal variable, possibly leading to
avalanches of nodal destabilizations cascading across the lattice.
Figure 8.2 illustrates schematically this redistribution process, in a
situation where node j exceeds the stability threshold through the
addition of a forcing increment δF (left panel). The subsequent
redistribution (right panel) pushes node $j+1$ above the stability
threshold, which will lead to a new redistribution of a nodal
quantity αF_{j+1}^{n+1} to nodes j and $j+2$ at the next iteration,
restoring the lattice to stability.

Notwithstanding the fact that it is defined here on a 2-D rather
than a 1-D lattice, the OFC model may look like a mere thematic
variation on the simple sandpile model introduced in chapter 5,
with the stability criterion defined in terms of the nodal values
themselves, rather than their slope (or gradient, in 2-D). The
apparently minor differences between the two model setups are
in fact profound at the level of their physical implications, and,
as we shall see in the remainder of the present chapter, lead to
markedly distinct global behaviors.

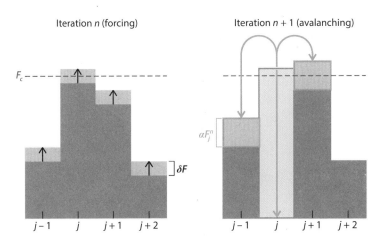

Figure 8.2. Action of the redistribution rule given by equations (8.12) and (8.13), here simplified to one spatial dimension. The lattice is everywhere stable ($F_j < F_c$) at the beginning of iteration n (left panel), but uniform forcing (black arrow; see equation (8.10)) pushes node j above the stability threshold F_c (dashed line). At the subsequent iteration (right panel), node j is reset to zero and only a fraction αF_j^n of its former value F_j^n is redistributed to neighboring nodes. Note that forcing stops during avalanching, i.e., this is a "stop-and-go" model. Compare to figure 5.1 for the 1-D sandpile model, where stability is based on the value of the slope.

One key difference is that for $\alpha < 0.25$ in equation (8.13), the OFC model is *nonconservative*: the sum of the nodal variables $F_{i,j}$ is smaller after a redistribution event than it was prior to it. Recall that the choice $\alpha = 0$ corresponds to a complete decoupling of the blocks with one another (the spring constant $K = 0$), in which case the force $F_{i,j}^n$ at an unstable node is entirely transferred to the upper plate though the leaf spring. It is only at the opposite extreme $\alpha = 0.25$, implying a ratio of spring constants $K_L/K \ll 1$, that all of $F_{i,j}^n$ is transmitted only to neighboring blocks during a slippage event, which then makes redistribution conservative.

Another key difference is that the driving, stability, and redistribution rules of the OFC model are all completely deterministic. The only stochasticity is introduced in the initial condition, where at $n = 0$ the nodal variable is set to some uniformly distributed random value within the allowed range of stable values:

$$F_{i,j}^0 = r, \quad i, j = 0, \ldots, N-1, \quad r \in [0, F_c]. \quad (8.14)$$

The OFC sandpile model is usually taken to operate in stop-and-go mode, meaning that driving is interrupted during avalanches and resumes only once the system is everywhere stable. The implied separation of timescales between the driving and avalanching processes is very well justified in the earthquake context, with mean displacement speeds for tectonic plates of about a centimeter per year (roughly the speed at which our nails grow), versus meters per second for slippage during earthquakes.

8.2 Numerical Implementation

The numerical implementation of the OFC model used in what follows, as listed in figure 8.3, closely follows that of the 1-D sandpile model in terms of overall code structure:

1. The simulation executes a preset number of temporal iterations, as set by the value of the variable `n_iter` (loop starting on line 20).
2. Once again that stability check and redistribution are executed one after the other, within the outer temporal loop, so as to achieve synchronous update of the nodal variable.
3. The lattice arrays `force` and `move` are assigned sizes $(N+2) \times (N+2)$, even though the lattice is of size $N \times N$ (lines 13 and 21). The extra rows and columns are ghosts nodes along the perimeter of the lattice, introduced so as to avoid out-of-bounds indexing

```
 1 | # OLAMI-CHRISTENSEN-FEDER 2D LATTICE MODEL FOR EARTHQUAKES
 2 | import numpy as np
 3 | import matplotlib.pyplot as plt
 4 | #-------------------------------------------------------------------------
 5 | N       =64                       # lattice size
 6 | f_thresh=5.                       # force threshold
 7 | delta_f =1.e-4                    # forcing amplitude
 8 | alpha   =0.15                     # conservation parameter
 9 | n_iter  =100000                   # number of temporal iterations
10 | #-------------------------------------------------------------------------
11 | dx=np.array([-1,0,1,0])           # template arrays
12 | dy=np.array([0,-1,0,1])           # template arrays
13 | force=np.zeros([N+2,N+2])         # force array
14 | toppling=np.zeros(n_iter,dtype='int')    # toppling time series
15 | totalf=np.zeros(n_iter,dtype='int')      # total force time series
16 | for i in range(1,N+1):
17 |     for j in range(1,N+1):
18 |         force[i,j]=f_thresh*(np.random.uniform()) # random initial force
19 |
20 | for iterate in range(0,n_iter):           # temporal iteration
21 |     move =np.zeros([N+2,N+2])             # reset evolution array
22 |     # scan lattice to flag which nodes must redistribute and reset to zero
23 |     for i in range(1,N+1):
24 |         for j in range(1,N+1):
25 |             if force[i,j] >= f_thresh:    # node i,j is unstable
26 |                 move[i,j]-=force[i,j]             # Eq (8.13): reset
27 |                 move[i+dx[:],j+dy[:]]+=alpha*force[i,j] # Eq (8.14): displace
28 |                 toppling[iterate]+=1      # cumulate topplings
29 |     # end of lattice scan
30 |
31 |     if toppling[iterate] > 0:             # avalanche occured
32 |         force+=move                       # update lattice
33 |     else:                                 # no avalanche
34 |         force[:,:]+=delta_f               # Eq (8.11): drive lattice
35 |
36 |     totalf[iterate]=force.sum()           # total force on lattice
37 |     if iterate % 10000 == 0:
38 |         print("{0}, toppl {1}.".format(iterate,toppling[iterate]))
39 | # end of temporal loop
40 | plt.subplot(2,1,1)
41 | plt.plot(range(0,n_iter),totalf)          # plot total force time series
42 | plt.ylabel('Total force')
43 | plt.subplot(2,1,2)
44 | plt.plot(range(0,n_iter),toppling)        # plot toppling time series
45 | plt.xlabel('Iteration')
46 | plt.ylabel('Toppling nodes')
47 | plt.show()
48 | # END
```

Figure 8.3. Minimal Python source code for the Olami–Feder–Christensen lattice-based implementation of the Burridge–Knopoff earthquake model.

(index $<$ 0 or $>$ N $-$ 1) during redistribution. The lattice loops therefore run from from index values 1 to N (lines 16–17 and 23–24). The forces on ghost nodes retain a value of zero throughout the simulation.

4. As with the forest-fire code of figure 6.1, two integer arrays, dx and dy, are used to define a nearest-neighbor template relative to any node (i, j) (lines 11–12); implementing equation (8.13) is then carried out via an implicit loop over the elements of these template arrays, by using them to index the move array (line 27).

5. Forcing takes place at all nodes (line 34), but only if no node was found to be unstable at the current iteration.

As with most Python codes introduced in the preceding chapters, this implementation favors readability over computational efficiency. Since the forcing is deterministic, the current state of the (non-avalanching) lattice determines entirely how many forcing iterations are required before the next toppling occurs; taking advantage of this fact can lead to significant speedup, the more so the smaller the δF. One of the computational exploration exercises at the end of this chapter offers a few hints on how to take advantage of this property of the OFC model.

8.3 A Representative Simulation

As usual, we first examine in some detail one specific representative simulation, here on a 128×128 lattice, with parameter values $F_c = 1$, $\delta F = 10^{-4}$, and $\alpha = 0.15$, the latter implying markedly nonconservative redistribution, as 40% of the nodal variable is "lost" every time a node topples. A good measure of avalanche size E here is the amount of force dissipated in the course of all redistribution events occurring during the avalanche. In practice, an equivalent measure is simply the total number of toppling nodes (counting all repeated toppling as such), since all redistribution events dissipate essentially the same quantity of the nodal variable, namely, $(1 - 4\alpha) \times F_c$, provided $\delta F/F_c \ll 1$.

Figure 8.4 shows portions of the avalanche-size time series for our representative simulation, after it has reached its statistically stationary state. The top panel shows a 40,000 iteration-long segment, the middle panel 2×10^5 iterations, then up to 2×10^6 at the bottom. Avalanches covering a wide variety of sizes are seen to occur, here ranging from 1 toppling up to 4000 for the largest avalanche in the bottom panel. The top time series shows a very clear recurrence of the same avalanching pattern, with period here $\simeq 10{,}960$ iterations. Careful examination of the time series reveals that it is not exactly periodic, with changes in the temporal patterns of the smaller avalanches. Going to longer time spans (middle and bottom panels) reveals that episodes of almost-periodic behavior have a finite temporal duration, gradually transiting from one recurrent avalanching pattern to another. The middle panel shows one such transition, in which the largest avalanche in the recurring pattern goes from a size of 750 in the first third of the sequence, up to 1800 in its final third. Nonetheless, over much longer time spans (bottom panel) there is clear periodicity present in the recurrence of the largest avalanches, despite large variations in their peak sizes during quasiperiodic subintervals.

Figure 8.5 shows time series of the nodal force $F_{i,j}$ at three selected nodes in the lattice's interior. The recurrence cycle is now strikingly apparent. Nodal values rise slowly, at the same rate for all nodes, in response to forcing, but these slow rises are interrupted by upward jumps by a quantity $\alpha F_c = 0.15$ here, when the node receives a force increment from a neighboring avalanching node, and drops to zero when the node itself exceeds the stability threshold. Here the blue and green nodes topple in response to an avalanching nearest neighbor, while the red node reaches the stability threshold via forcing. The colored line segments in the top panel of figure 8.4 indicate the times when

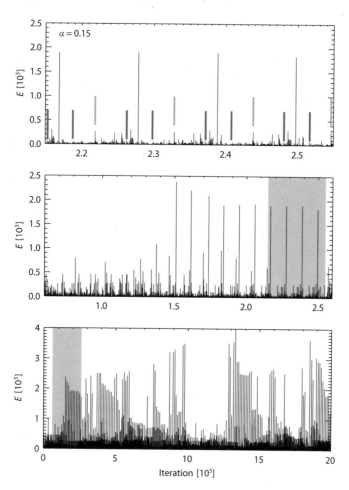

Figure 8.4. Three segments of increasing lengths extracted from the time series of avalanche size in a simulation executed on a 128×128 lattice, with conservation parameter $\alpha = 0.15$ and forcing parameter $\delta F = 10^{-4}$. The top panel spans 4×10^4 iterations, the middle 5 times more, and the bottom another factor of 10 more. The shaded areas indicate the temporal range covered by the preceding panel. The colored line segments in the top panel indicate the toppling times for three selected lattice nodes (see figure 8.5). The avalanche-energy time series exhibit clear periodic behavior, here with a period of $\simeq 10,960$ iterations.

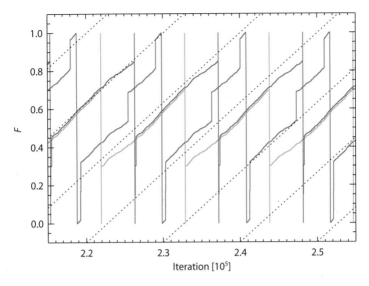

Figure 8.5. Time series of the force value at three selected nodes on the lattice, in the same simulation as in figure 8.4 (128×128, $F_c = 1$, $\delta F = 10^{-4}$, $\alpha = 0.15$). These three time series span the same interval as the top panel in figure 8.4. The sampled nodes are $(i, j) = (64, 64)$, $(32, 32)$, and $(64, 76)$, shown in red, green, and blue, respectively. The slanted dotted lines indicate a growth rate of $\delta F = 10^{-4}$ per iteration, corrected for the mean fraction of iterations spent avalanching.

the three sample nodes of figure 8.5 are avalanching. None of the three nodes participate in the largest periodic avalanches, but the blue and green nodes do take part in smaller recurrent avalanches still large enough to be distinguishable on the scale of this plot.

The recurrence period of $\simeq 10{,}960$ iterations is conspicuously close to the time $t = (\delta F)^{-1} = 10^4$ iterations required for forcing alone to take a node from zero up to the stability threshold. This is only part of the story though, because all nodes, in the course of their buildup, jump up a few times in response to an avalanching neighbor. Moreover, the model is operating in stop-and-go mode under a single temporal iteration loop; iterations spent avalanching must be subtracted from the recurrence period if comparing it to the forcing timescale. In this specific simulation, 63% of iterations are spent avalanching somewhere on the lattice, so that the "corrected" recurrence period measured in forcing steps is in fact 4000 iterations, leading to a growth in $F_{i,j}$ of 0.4 under pure forcing. The remainder is produced by avalanching neighbors, consistent with figure 8.5. Put differently, in the course of a recurrence cycle, an "average" node receives four increments of $+0.15$ from an avalanching neighbor, and the rest from the deterministic driving process, consistent with figure 8.5.

We are still faced with a puzzle: how can a purely deterministic evolution using a totally random initial condition produce (quasi)periodic global behavior on (relatively) short timescales, but aperiodic on longer timescales? The nodal coupling mediated by the redistribution rule is the culprit. Figure 8.6 shows the lattice initial condition (top left), and also three times, separated by 10^6 iterations, much later in the simulation, as labeled. The random initial pattern gives way to a patchwork of domains of varying sizes, within which many nodes have the same value or share a small set of values. These domains vary in shape and size as the simulation unfolds, as they interact with one another through avalanching taking place at their boundaries. These spatial domains of contiguous, similar nodal values are directly reflected in the recurrent avalanching patterns of figure 8.4. Whether a domain is destabilized through a neighboring avalanche or because all nodes hit the stability threshold at the same time through slow forcing, the whole domain collapses to zero and rebuilds anew. The larger the domain, the larger the associated avalanche. We have encountered something like this already with the forest-fire model of chapter 6 (cf. figure 6.8). The slow evolution of domain sizes and boundaries is what leads to the gradual transitions between different recurrent avalanching patterns, as exemplified by the middle panel in figure 8.4.

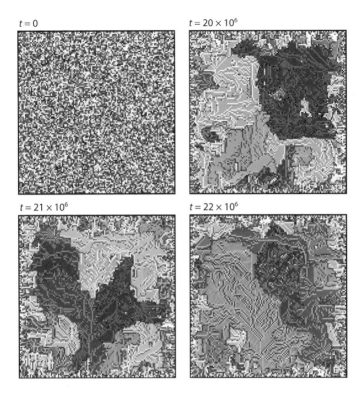

Figure 8.6. Four snapshots of the OFC lattice, for the same simulation as in figures 8.4 and 8.5 ($\delta F = 10^{-4}$, $\alpha = 0.15$). The color scale encodes the magnitude of the force $F_{i,j}$ at each node. The top-left frame shows the purely random initial condition ($F_{i,j}$ randomly distributed in the interval [0, 1]), and the other three frames are sampled at a cadence of 10^6 iterations, as labeled. Note the buildup of large "domains" of equal values in F_{ij}, slowly changing shape as the simulation proceeds, and compare with figure 6.8.

8.4 Model Behavior

There are three parameters defining model behavior: the conservation parameter α, forcing parameter δF, and threshold F_c. At a given α, all that matters is the ratio $\delta F / F_c$; hereafter we continue to use $F_c = 1$, as in the representative simulation just considered, without loss of generality. The choice of δF is largely irrelevant to the avalanching dynamics, as long as it is small enough, in the sense of $\delta F \ll \alpha F_c$, corresponding to slow forcing. The adopted value of δF does set the mean inter-avalanche waiting time, though, and therefore the overall timescale of the simulation.

Running the OFC model with different values of α soon reveals that the numerical value of this parameter has an important influence on the size and recurrence period of avalanches. In the limiting case of no internodal coupling ($\alpha = 0$), each nodal value grows linearly at a rate (δF) per iteration, starting from its (random) initial condition, and subsequently avalanches independently of neighboring nodes at exactly every $(\delta F)^{-1}$ iterations. The system is completely periodic, all avalanches are of size 1 (unless two nodes have initial values that differ by less than δF), and the random initial pattern is forever frozen into the system. This is no longer the case when redistribution couples avalanching nodes ($\alpha > 0$). As α increases, the recurrence period diminishes, from 6435 iterations at $\alpha = 1.0$, 4002 at $\alpha = 1.5$, down to 2165 iterations at $\alpha = 0.2$. This trend makes sense in light of our earlier discussion of figure 8.5. The higher α, the larger the upward jump in nodal value in response to an avalanching neighbor. Correspondingly fewer forcing iterations are then required to reach the stability threshold. The recurrent avalanching patterns disappear gradually as $\alpha \to 0.25$, and are nowhere to be found at $\alpha = 0.25$

Whatever the value of α, periodic behavior is due to the presence of spatial subdomains of identical nodal values on the lattice (see figure 8.6). It is easy to understand how a large domain of contiguous nodes sharing the same nodal value will remain "synchronized" over extended periods of time. How that synchronization sets in, starting from a purely random initial pattern, is what begs an explanation.

Consider two neighboring nodes $F_{(1)}$, $F_{(2)}$ with force values

$$F_{(1)}^n = \bar{F} + \Delta,$$
$$F_{(2)}^n = \bar{F} - \Delta \rightarrow |F_{(2)}^n - F_{(1)}^n| = 2\Delta. \tag{8.15}$$

Now suppose that both nodes are avalanching simultaneously; as per the redistribution rules (8.12) and our synchronous nodal updating procedure, their post-redistribution value will not be $F_{(1)} = F_{(2)} = 0$, but rather

$$F_{(1)}^{n+1} = \alpha(\bar{F} - \Delta),$$
$$F_{(2)}^{n+1} = \alpha(\bar{F} + \Delta) \rightarrow |F_{(2)}^{n+1} - F_{(1)}^{n+1}| = 2\alpha\Delta, \tag{8.16}$$

assuming again here that $F_c = 1$, without loss of generality. The difference in nodal values prior to and after the redistribution has thus decreased, by a factor α (≤ 0.25).[4] Once the two nodes are synchronized (in the sense $F_{(1)}^n = F_{(2)}^n$, so that $\Delta = 0$ in equation (8.15)), redistribution will maintain synchrony, and so will the deterministic forcing mechanism embodied in equation (8.10); only the input from a third neighboring avalanching node can break it. In other words, once a spatially extended portion of the lattice is synchronized, it can only be destroyed at its boundary; the more extended the synchronized region, the longer it is likely to persist, as boundary perturbations make their way inward in successive recurrence cycles.

It is a remarkable fact that despite the model's (quasi)periodic temporal behavior, avalanches in the OFC model remain scale invariant. This is illustrated in figure 8.7, showing PDFs of avalanche sizes for simulations using $\alpha = 0.1, 0.2$, and 0.25. The latter is conservative, and is characterized by a power law spanning

---

[4]This is typical of isotropic linear diffusive processes, which tend to even out gradients in the diffusing quantity. Indeed, here the net quantity of "force" transported from the higher- to the lower-valued node by the redistribution rule is $(1 - \alpha)2\Delta$, which is linearly proportional to the initial difference 2Δ in the nodal value, in line with classical Fickian diffusion.

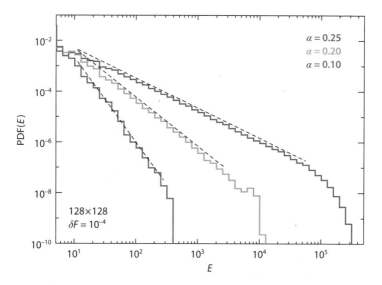

Figure 8.7. Probability density functions of avalanche energy in simulations with varying values of the conservation parameter α, as color coded. All distribution have a well-defined power-law range, with logarithmic slope flattening with the conservation parameter: -3.34 at $\alpha = 0.1$, -1.92 at $\alpha = 0.2$, up to -1.19 at $\alpha = 0.25$. The dotted line segments indicate the energy range over which these slopes are computed. All three simulations are executed on a 128×128 lattice, with forcing parameter $\delta F = 10^{-4}$. These distributions are based on 5×10^6-iteration-long segments, during which over 10^6 avalanches have taken place in each simulation.

over four orders of magnitude in avalanche size, with logarithmic slope -1.19. Nonconservative simulations ($\alpha < 0.25$) retain the power-law shape, with the logarithmic slope steepening and the upper cutoff moving to smaller sizes as α decreases. These trends are readily understood upon noting that, for a statistically stationary state to be maintained, the nodal variable must be either dissipated locally or evacuated at the boundaries, at the same average rate as the forcing increases it. Low levels of dissipation thus require more avalanches to discharge at the boundaries, while at higher levels of dissipation, avalanches can more easily stop somewhere in the interior of the lattice. Consequently, large avalanches become more common as $\alpha \to 0.25$, which translates into a flatter power law for their size distribution.

8.5 Predicting Real Earthquakes

Large earthquakes are extremely destructive, either by themselves or through the tsunamis they can generate. They are, arguably, the one type of global natural hazard one would most like to be able to predict. Not surprisingly, seismologists and mathematicians, as well as a wide assortment of quacks, have been at it for years and years. In part because such a large volume of data is available, some earthquake prediction schemes have been proposed, based on purely statistical inferences or artificial intelligence-based expert systems. Such techniques typically strive to identify robust precursor signals in time series of seismic data. Much like the lattice-based model introduced in this chapter, these prediction techniques are based on highly simplified representations of the physical processes underlying the target phenomenon—if they are at all present in the model. In fact, they operate by *training* on real-world data, by learning to "recognize," in the data, patterns that have predictive value.

Consider now the consequences of earthquake magnitude being distributed as a power law, as per the Gutenberg–Richter law. Most events are small, and the larger events, which are those one would very much like to predict, are rare. Therefore the seismic data record, as voluminous as it may be, contains mostly small events. The number of large events available to train expert systems ends up being rather small, the more so the larger the target event size. This is (at least in part) why reliable earthquake prediction schemes are still lacking despite many decades of data collection and research efforts.

The Gutenberg–Richter law is characterized by logarithmic slope $b \sim -1$ for a cumulative distribution, implying $\simeq -2$ for PDFs such as plotted in figure 8.7. Taken at face value, the OFC model would then indicate that plate tectonics operates in the nonconservative regime, with α somewhere in the range 1.5–2.0. This is a parameter range where recurrent avalanching behavior sometimes occurs. This possibility is supported to some extent by seismic data, which show that certain tectonic faults, including the (in)famous San Andreas fault in California, generate large "characteristic" earthquakes which exhibit quasiperiodicity in their temporal pattern of occurrence.

Quasiperiodicity is an extremely attractive property from the point of view of earthquake forecasting, and you get to try your hand at it in the Grand Challenge for this chapter. But beware, this is a serious and dangerous business. A team of six Italian government seismologists found out the hard way when they were prosecuted and found guilty of manslaughter for having failed to predict the April 2009 earthquake that destroyed the small town of L'Aquila.[5] Bear in mind also that the OFC sandpile is a model of the Burridge–Knopoff model of seismic faults; when you're making predictions using a model of a model, caution is definitely warranted, even if you don't live in Italy.

8.6 Exercises and Further Computational Explorations

1. Fill in the missing mathematical steps leading from equation (8.3) to (8.8).
2. Compute the PDFs of avalanche sizes for simulations using $\alpha = 0.2$ and $\delta F = 10^{-4}$ on lattice sizes 32×32, 64×64, and 128×128. Compare the logarithmic slopes and large-size cutoffs. Be careful to build your

---

[5]Rational thinking—or perhaps just plain common sense—eventually prevailed, and the conviction was finally overturned in November 2014 by an appeal court.

PDFs from time series segments in the statistically stationary state; this may require up to 10^7 iterations for the larger lattice size (but see the next exercise)!

3. Knowing the value of the forcing parameter δF and the largest nodal value $F_{i,j}^n$ on the (non-avalanching) lattice at iteration n, one can easily compute the number of iterations required before the triggering of the next avalanche: $(F_c - \max(F_{i,j}^n))/\delta F$. This result can be capitalized upon to accelerate the simple-minded code of figure 8.3. Do it, and estimate the speedup factor.

4. Introduce into the model a mildly stochastic nonconservative redistribution by drawing anew a value of α uniformly distributed in the range $0.14 \leq \alpha \leq 0.16$ at each toppling node. Is this enough to break the quasiperiodicities of the avalanche time series?

5. Construct correlation plots between avalanche sizes (E) and duration (T) for a set of three simulations using $\alpha = 0.15, 0.2$, and 0.25. Can you infer a mathematical equation that captures the (statistical) relationship between these two quantities?

6. And now for the Grand Challenge: earthquake prediction! Extract a 2×10^5-iteration-long segment of the avalanche time series in the statistically stationary state of an $\alpha = 0.15$, $\delta F = 10^{-4}$ simulation. Using the first 10^5 iterations, compute the maximum avalanche size, the recurrence period of avalanches, and whatever other potentially useful quantities you may think of. Then, try to forecast the timing and size of the larger avalanches (size larger than 20% of the maximum avalanche size determined previously) in the second half of your time series. Here a "good forecast" means getting the timing of the earthquake right to within 10^2 iterations, and amplitude within to $\pm 25\%$ of the

"observed" value. Keep track also of *false alarms*, when you predict a large earthquake that does not occur, and *misses*, when you fail to predict an earthquake that does occur. This is a pretty open-ended Grand Challenge; you get to decide when to stop!

8.7 Further Reading

I am no expert on earthquakes, which is perhaps why I found the Wikipedia page on the topic informative and a very good read (consulted December 2014):

http://en.wikipedia.org/wiki/Earthquake.

The web page of the US Geological Survey also contains a wealth of interesting information on earthquakes, and provides access to all kinds of earthquake-related data:

http://earthquake.usgs.gov/earthquakes/.

On the Burridge–Knopoff model and its sandpile-like reformulation, see

Carlson, J.M., and Langer, J.S., "Mechanical model of an earthquake fault," *Phys. Rev. A*, **40**, 6470 (1989);

Olami, Z., Feder, H.J.S., and Christensen, K., "Self-organized criticality in a continuous, nonconservative cellular automaton modeling earthquakes," *Phys. Rev. Lett.*, **68**, 1244 (1992);

Hergarten, S., *Self-Organized Criticality in Earth Systems*, Berlin: Springer, chap. 7 (2002).

The description of the Burridge–Knopoff model in section 8.1 is closely inspired by the presentation in section 3.10 of the following book, which also offers an illuminating mathematical analysis of the OFC model:

Christensen, K., and Moloney, N.R., *Complexity and Criticality*, Imperial College Press (2005).

On more elaborate modeling and analyses of earthquakes, a good recent entry point in the literature is

Sachs, M.K., Rundle, J.B., Holiday, J.R., Gran, J., Yoder, M., Turcotte, D.L., and Graves, W., "Self-organizing complex earthquakes: Scaling in data, models and forecasting," in *Self-Organized Criticality Systems*, ed. M.J. Aschwanden, Berlin: Open Academic Press, 333–356 (2013).

9

EPIDEMICS

Whether the Black Death in the Middle Ages, AIDS in the 1980s, or the 2014 Ebola outbreak in Africa, all epidemics are *scary* (like earthquakes)! Perhaps it is the thought that even the strongest among us can be felled, almost randomly, by an organism that cannot even be seen or felt; or the fact that huddling together, a most natural human reflex in times of duress, is exactly what we should *not* be doing during an epidemic.

It turns out that the epidemic spread of contagious diseases shares many characteristics with some of the apparently unrelated systems considered in preceding chapters. Let's dive in and look into that.

9.1 Model Definition

Contagious diseases are often said to spread "like wildfires," and this is precisely the basic idea underlying the model of epidemic spread considered in this chapter. More specifically, the model is constructed by adding random walks on a lattice to the forest-fire model of chapter 6. The algorithm is as follows: a preset number M of random-walking agents move on a 2-D Cartesian lattice of size $N \times N$, with two or more agents allowed to meet on the same lattice node. A contagious agent is now introduced at some random location on the lattice. Perhaps fortunately for himself (but certainly not for the remainder of the population), the sick agent does not keel over immediately, but survives during L temporal iterations, and infects any other random-walking agent met on any lattice node during that time period. These newly infected agents immediately become contagious and also have a postinfection life span of L iterations, and so will likely also infect other agents met on other lattice nodes in the subsequent L iterations. Sick walkers fall dead on the spot L iterations after being infected, remain immobile thereafter (no zombies!), and immediately cease to be contagious.

As you may well imagine, this algorithm can lead to an "avalanche" of infection events propagating across the lattice; but, as usual, looking into how this comes about will reveal some interesting subtleties, and complexities!

9.2 Numerical Implementation

Figure 9.1 is a minimal Python source code implementing the epidemic "algorithm" just described. Take good note of the following:

1. The simulation is structured within an outer conditional (`while`) loop (starting on line 24), which will iterate until the number of infected walkers (`n_sick`) falls to zero, or a preset maximum number of temporal iterations (`max_iter`) has been reached. Upon termination of this loop, the value of the variable `iterate` yields the duration of the epidemic.

2. Four 1-D arrays of length M store the information characterizing each random walker: `x[M]`, `y[M]`, `infect[M]`, and `lifespan[M]` (lines 12–14). The (x, y) position on the lattice of the jth walker is stored in elements `x[j]` and `y[j]`; its medical status is stored in `infect[j]`, where a value 0 indicates a healthy walker, 1 an infected walker, and 2 a deceased,

```
1   # EPIDEMIC SPREAD IN A POPULATION OF RANDOM WALKERS ON A LATTICE
2   import numpy as np
3   import matplotlib.pyplot as plt
4   #------------------------------------------------------------------------
5   N       =128                       # lattice size
6   M       =4000                      # number of random walkers
7   L       =20                        # lifetime parameter
8   max_iter=10000                     # maximum number of iterations
9   #------------------------------------------------------------------------
10  x_step  =np.array([-1,0,1,0])      # template arrays
11  y_step  =np.array([0,-1,0,1])
12  x,y     =np.zeros(M),np.zeros(M)   # walker (x,y) coordinates
13  infect  =np.zeros(M)               # walker health status
14  lifespan=np.zeros(M)               # time left to live
15  ts_sick =np.zeros(max_iter)        # time series of sick walkers
16  for j in range(M):                 # place walkers on lattice
17      x[j]=np.random.random_integers(0,N)
18      y[j]=np.random.random_integers(0,N)
19      lifespan[j]=L
20  jj=np.random.random_integers(0,M-1)   # infect one random walker
21  infect[jj]=1
22  n_sick,n_dead,iterate=1,0,0        # various counters
23
24  while (n_sick > 0) and (iterate < max_iter):  # temporal iteration
25      for j in range(0,M):              # loop over all walkers
26          if infect[j] < 2:             # this walker is still alive
27              ii=np.random.choice([0,1,2,3])  # pick direction
28              x[j]+=x_step[ii]          # update walker coordinates
29              y[j]+=y_step[ii]
30              x[j]=min(N,max(x[j],1))   # bounding walls in x,y
31              y[j]=min(N,max(y[j],1))
32              if infect[j]==1:          # this walker is sick
33                  lifespan[j]-=1        # the clock ticks...
34                  if lifespan[j] <= 0:  # this walker dies
35                      infect[j]=2
36                      n_sick-=1
37                      n_dead+=1
38              for k in range(0,M):      # check for walkers on node
39                  if infect[k] == 0 and k != j: # this one is healthy...
40                      if x[j]==x[k] and y[j]==y[k]:   # ...and so: infect
41                          infect[k]=1
42                          n_sick+=1
43      # end of loop over all walkers
44      ts_sick[iterate]=n_sick
45      iterate+=1
46      print("iteration {0}, sick {1}, dead {2}.".format(iterate,n_sick,n_dead))
47  # end of temporal loop
48  plt.plot(range(0,iterate+10),ts_sick[0:iterate+10])
49  plt.show()
50  # END
```

Figure 9.1. Minimal Python source code for the epidemic-spread model used in this chapter.

immobile walker. The element `lifespan[j]` is the life span of the walker, i.e., the number of temporal iterations it has left to live.

3. Initialization consists in randomly distributing the M walkers on the lattice (lines 17–18), and assigning infected status (`infect[jj]=1`) to a randomly chosen single one (lines 20–21).

4. Only the live walkers (whether healthy or infected, `infect[j]<2`) move on the lattice (line 26). The 2-D random walk is done in the usual manner (lines 27–29), but here a combination of min/max operations ensure that walkers on edge nodes cannot leave the lattice (lines 30–31). Put otherwise, the walkers are trapped on a square-shaped island. Are we starting to feel nervous?

5. Once walker j becomes infected, the corresponding array element `lifespan[j]` is decremented by 1 at each subsequent temporal iteration (line 33); when `lifespan[j]` hits zero (line 34), walker j is declared dead (`infect[j]=2`; second `if` statement in the inner loop).

6. For each infected walker (last `if` statement in the walker loop), another inner loop (starting on line 38) tests for coincidence in the x and y lattice coordinates with any other healthy (`infect[k]==0`) walker (line 39), in which case infection occurs (`infect[k]=1` on line 41).

7. Counter variables `n_sick` and `n_dead` accumulate the number of infected and dead walkers on the lattice at each temporal iteration, and are written to the screen at each iteration.

8. Upon termination of the epidemic, a time series of the number of sick walkers (previously accumulated in array `ts_sick`) is plotted against the iteration count, like the red curve in figure 9.2.

Here again, it would be possible to take advantage of the operators and functions for list manipulation provided by Python in order to design a computationally more efficient version of this simulation. One of the exercises at the end of this chapter offers hints on how to get started.

9.3 A Representative Simulation

As is now customary, we first look in some detail at the characteristics of a specific simulation, before launching into a study of the model's behavior. Figure 9.2 shows a time series of the number of infected (red line) and healthy (green line) individual walkers in a simulation beginning with $M = 4000$ random walkers moving on a 128×128 lattice, including one randomly selected, infected individual. The lifespan parameter is set here at $L = 20$ temporal iterations. Even though the epidemic begins with a single infected individual, it spreads here rather rapidly, with some 50 individuals already infected by the time the original sick individual keels over at iteration 20. By iteration 100, the epidemic looks like it is waning, but it then picks up again, in fits and starts, to reach its peak with 73 infected individuals at iteration 262. This is followed by five more or less distinct epidemic surges, before the last infected individual finally dies at iteration 919, marking the end of this simulated epidemic.

Figure 9.3 shows the spatiotemporal spread of the epidemic, this time by plotting a color-coded symbol at the final resting place of each infected individual in the simulation. The color code, indicated in the color bar, gives the iteration at which each individual was infected. The line segments connect each infected individual with the one having transmitted the disease to them; note that on this plot these are at distinct spatial locations, since individuals are not plotted at the locations where they were infected. Here the original infected individual is located very near the lower edge of the lattice (indicated by the orange circle), thus with fewer walkers within its 20 iteration range than if it

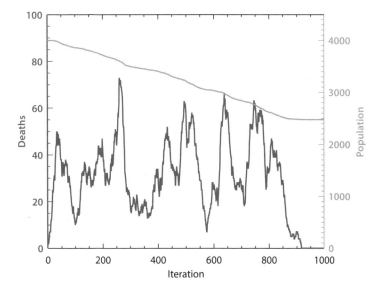

Figure 9.2. Time series of infected (red) and healthy (green) random walkers in a simulation carried out on a 128 × 128 lattice initially populated by 4000 random walkers, with the life span of an infected walker set at $L = 20$ iterations. Here, the duration of the epidemic is 919 temporal iterations, significantly above average for these parameter settings; yet, at the end of the epidemic, the population is reduced to 62% of its initial value, which is almost exactly the average death toll for these parameter settings (more on all of this in section 9.4). Note the multiple successive surges in the evolution of the epidemic, a characteristic commonly observed in real epidemics, and compare to figure 6.3.

had been located in the interior. Nonetheless, in this specific instance, the epidemic does manage to spread, initially in more or less all directions. The resulting circular "epidemic front" first increases in radius, but soon breaks into two distinct subfronts, one heading vertically upward and the other meandering to the right and eventually extinguishing in the bottom-right corner of the lattice. The first front fares better (so to speak), as it

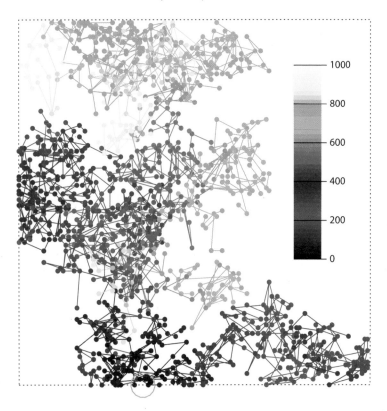

Figure 9.3. Epidemic spread for a simulation on a 128 × 128 lattice populated by 4000 random walkers, with the lifetime of an infected walker set at $L = 20$ iterations (same simulation as in figure 9.2). The solid dots indicate the final resting place of dead walkers, color coded according to the iteration at which they became infected, as indicated by the color bar, and the line segments connect each dead walker to the walker from whom the infection was picked up. The simulation was initialized with the introduction of a single infected walker, here very near the bottom edge of the lattice, as indicated by the orange circle. In this simulation, 1534 walkers fell to the epidemic, very close to the average for this initial population density and lifetime parameter.

spreads upward to the top of the lattice, with yet another subfront backtracking downward to end very near the point of origin of the epidemic. Here the epidemic has thus "percolated" from one end of the lattice to the other, so that this type of process is akin to *dynamical percolation*.

Qualitatively speaking, the epidemic surges on the time series of figure 9.2 and the spatially distinct infection foci in figure 9.3 are both features observed in real epidemics. Because no new random walkers are introduced into the lattice (ruthless quarantine in effect!), the epidemic inevitably ends as it destroys its own propagation vector. In the case of the simulation displayed in figures 9.2 and 9.3, the epidemic ends after 919 iterations, with over a third of the initial population killed off; we are in the general ballpark of the Black Death here, with an estimated 25% of Europe's population wiped out between 1347 and 1350.

9.4 Model Behavior

It doesn't take much profound thinking to realize that a key factor in epidemic spread is the density of random walkers moving about on the lattice. The higher that density, the more likely an infected agent is to meet and infect at least one healthy colleague before dropping dead, and in doing so, sustain the epidemic. The initial population density (ρ) is simply defined as the ratio of the M walkers to the number of available lattice nodes ($N \times N$), and equivalent to an occupation probability,

$$\rho = \frac{M}{N^2}. \qquad (9.1)$$

Figure 9.4 shows results from a series of epidemic simulations, all with a lifespan parameter $L = 20$, and computed for varying initial population density. These are again ensemble average results; for each value of ρ, K statistically independent realizations are carried out, by changing the seed of the random number generator controlling the distribution of the initial population

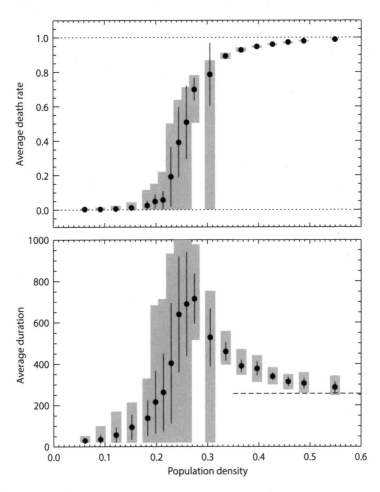

Figure 9.4. Variation of the death rate (top) and epidemic duration
(bottom), as a function of initial population density. Each solid dot
corresponds to the mean of 20 statistically independent realizations
of each simulation, with the vertical line segment indicating ± 1
standard deviations about that mean (see equations (9.3)). For each
density value, the gray bands indicate the minimum and maximum
values occurring in this 20 member set. All simulations have been
executed on a 128×128 lattice, with the lifespan of an infected
walker set at $L = 20$ iterations. Compare the top panel here to the
bottom panel of figure 4.6.

and random-walk steps. For each individual simulation, one can compute the duration T_k and death rate $0 \leq \mu_k \leq 1$, the latter being simply the ratio of deaths to the initial population size. The solid dots plotted in figure 9.4 are the values of μ and T averaged over each ensemble of $K = 20$ simulations at a given value of ρ, denoted in what follows as $\langle T \rangle$ and $\langle \mu \rangle$:

$$\langle \mu \rangle = \frac{1}{K} \sum_{k=1}^{K} \mu_k, \qquad \langle T \rangle = \frac{1}{K} \sum_{k=1}^{K} T_k. \qquad (9.2)$$

A quantitative measure of the variability in epidemic spread is offered by the root-mean-squared standard deviations about these ensemble averages:

$$\sigma_\mu = \left(\frac{1}{K} \sum_{k=1}^{K} (\mu_k - \langle \mu \rangle)^2 \right)^{1/2},$$

$$\sigma_T = \left(\frac{1}{K} \sum_{k=1}^{K} (T_k - \langle T \rangle)^2 \right)^{1/2}. \qquad (9.3)$$

The vertical line segments in figure 9.4 are drawn over the range $\langle \mu \rangle \pm \sigma_\mu$ and $\langle T \rangle \pm \sigma_T$, at each value of ρ.

Figure 9.4 certainly indicates that the average death rate $\langle \mu \rangle$ increases with population density, as expected, but the form of the variation should remind you of the growth of the largest cluster on a percolation lattice (cf. figure 4.6, bottom panel). As with the percolation problem, the variance is also largest at values of ρ where the average death rate varies most rapidly. This large variability also carries over to epidemic duration, which also peaks at the value of ρ around which the death rate increases the fastest. The subsequent decrease of the epidemic duration reflects the fact that even at very high population density, the model design is such that the epidemic front can advance at most by one nodal spacing per temporal iteration. This leads to a saturation of the epidemic duration at $\simeq 2 \times N = 256$, indicated by the dashed line segment in the bottom panel of figure 9.4; this is the total number of steps required on a 4-neighbor Cartesian lattice to travel from one corner of the lattice to the opposite corner.

Examination of the high variability around $\rho = 0.25$ soon reveals that the high rms deviation is a consequence of the epidemic simply failing to pick up in a subset of the simulations. In other words, the distribution of death rates or duration does not look at all like a Gaussian centered on the mean value— yet another reminder of the potential interpretative pitfalls of equations (9.3). The vertical gray bands in figure 9.4 span the range going from the lowest to the highest values of death rate and duration, in each 20 member set of simulations at each initial density. For low ($\rho \lesssim 0.15$) and high ($\rho \gtrsim 0.35$) density, this range is quite narrow and well centered on the mean value; but in between a much wider span is observed. Typically, in the range $0.2 \lesssim \rho \lesssim 0.3$, a given epidemic can fail to take off altogether, due to the injected sick agent keeling over after $L = 20$ iterations, before infecting anyone, or persist for many hundreds of iterations and achieve a high death rate. Even at the relatively high density of $\rho = 0.305$, here one individual simulation has failed to produce an epidemic, while seven of the other realizations eradicate over half the initial population. Evidently, what happens very early in the simulation is crucial.

At the beginning of the simulation, with a mean density $\rho = M/N^2$ and all walkers moving independently of one another, the probability of two walkers meeting on the same node at one iteration is given by ρ^2. If one walker is infected, the probability that it does *not* meet a healthy walker in L iterations is thus $(1 - \rho^2)^L$; therefore, the probability p that the infected agent meets at least one healthy member of the population is

$$p = 1 - (1 - \rho^2)^L. \tag{9.4}$$

For the $L = 20$ simulations used to construct figure 9.4, we have $p = 0.18$ at $\rho = 0.1$, rising to $p = 0.56$ already at $\rho = 0.2$, $p = 0.85$ at $\rho = 0.3$, up to $p = 0.97$ at $\rho = 0.4$. This indicates that the odds of the epidemic getting going exceed 50–50, once the initial density reaches $\simeq 0.2$. Between $\rho = 0.2$ and 0.3, there is still a fair chance that early on, the first or first few infected agents die before infecting others in the population, but when the epidemic does get going, a large number of agents will end up being infected. One can thus expect greater variability in the epidemic duration and death toll in this range. Beyond $\rho = 0.3$, the spread of infection is almost certain, and large epidemics invariably ensue. These probabilistic inferences are in general agreement with the numerical results of figure 9.4.

The above analysis is limited by the fact that in reality a sick walker can infect more than one healthy individual. For example, in the specific simulation of figures 9.2 and 9.3, the average infection rate is almost exactly one per infected walker, but a large fraction (40%) of infected walkers actually died without transmitting the disease, 34% of sick walkers infected only one population member, 17% managed to infect two, and so on, with the two most "efficient" infectors each managing to transmit the disease to six healthy population members; once again, nothing like a Gaussian distribution.

Ultimately, for the epidemic to grow, the total infection rate must exceed the death rate, but how many healthy walkers will be infected by a given sick walker depends on the local density of healthy walkers, itself influenced by the prior presence or absence of infected walkers in the vicinity. Figure 9.5 displays the epidemic spread for the simulation of figure 9.3 (on the left), as well as for a higher initial density simulation ($\rho = 0.49$, on the right). Now only the "infection links" between infected and infector are plotted, this time as a function of the x-nodal position of the walkers' final resting place, with time running vertically upward. The branching structure of the epidemic spread is now clearly visible, and, especially for the lower initial density simulation, has a definite self-similar look. One can also pick up a definite maximal inclination for the spreading branches

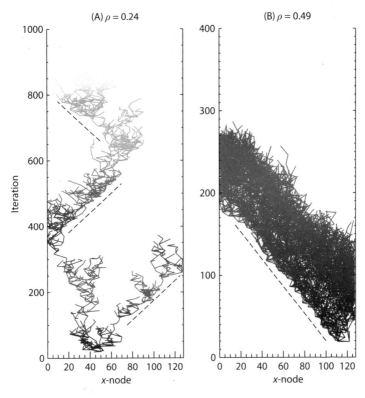

Figure 9.5. Another view of epidemic spread, now with time running vertically upward and the dead walkers distributed according to their horizontal nodal position on the lattice. Only the "infection links" are plotted, for clarity. Panel A is for the same simulation as in figure 9.3. Note the self-similar branching structure, and how this epidemic would have ended around iteration 380, had it not been for the single infection event taking place at $(x, t) = (18, 282)$. Panel B shows an epidemic spread in a simulation with twice the initial population density of panel A. Note the different vertical (temporal) scales on both panels. The dashed oblique lines are a guide to the eye, indicating propagation speeds of 0.33 nodes per iteration on the left, and 0.6 nodes per iteration on the right. The death rates for these simulations are 38.8% in panel A, and 98.2% in panel B.

in these space-time diagrams, indicative of a well-defined peak propagation speed. The dashed line segments are "guides to the eye," serving to indicate this peak propagation speed, which here increases almost linearly with initial population density.

The left panel in figure 9.5 illustrates well why simulations such as this one exhibit the most variability in figure 9.4. At this initial density, the epidemic is barely able to propagate across the lattice. Just as a single infection event starts the epidemic at the beginning of the simulation, here, much later in the simulation, often a single infection event determines whether the epidemic will extinguish itself or flare up again. This is the case with the $\rho = 0.24$ simulation of figure 9.5A, in which the single infection event having taken place at $(x, t) = (18, 282)$ is responsible for the further spread of the epidemic for another 650 iterations; had this infection not taken place, the epidemic would have ended with the extinction of its right branch, around iteration 380. Think about it: all it would have taken is one random step in a different direction.

The epidemic spread at the higher density $\rho = 0.49$ (right panel of figure 9.5) is more regular, as the epidemic front progresses steadily across the lattice, leaving only dead walkers in its wake. Figure 9.6 offers yet a different view of epidemic propagation at high density. The top panel is a series of snapshots, 20 iterations apart, showing the spatiotemporal evolution of the spatial density of sick agents. This is computed simply by dividing up the 128×128 lattice in 16×16 contiguous blocks of 8×8 nodes, and computing the number of sick agents in each such block. The resulting 16×16 array is then rendered in grayscale. The epidemic spreads as a more or less circular wave front, like ripples on the surface of a pond into which a rock has been thrown. For each such snapshot, one can compute the distance of each infected agent from the starting point of the epidemic, namely, the node on which the initial single infected agent was

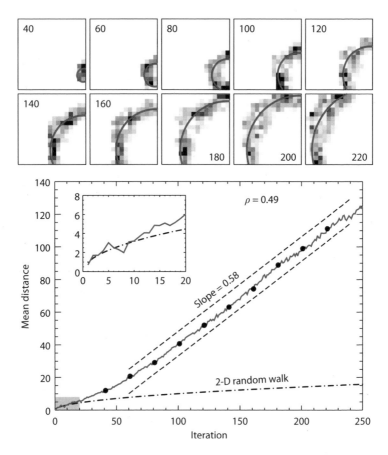

Figure 9.6. Spread of the epidemic wave front in the $\rho = 0.49$ solution of figure 9.5B. The sequence of snapshots at the top shows a grayscale rendering of the density of infected agents, at a 20-iteration cadence, as labeled. The red circular arcs are centered on the starting point of the epidemic, and drawn with the mean distance of the infected agents measured from that point. The bottom panel shows the variation with time of this distance (solid line), along with the root-mean-squared displacement for a 2-D random walk (dash-dotted line). The solid dots identify the 10 snapshots plotted at the top, and the inset is a close-up of the first 20 iterations.

located. The red circular arcs on the snapshots are centered on this location and drawn with a radius equal to the mean distance of all infected agents. The wave front thickens and develops internal structure (i.e., density "clumps"), but retains its circular shape until the lattice boundaries are encountered. The bottom plot shows how this mean radius varies with time. From about iteration 80 onward, the relationship is linear, indicating a radial expansion of the epidemic front at a constant speed of $\simeq 0.6$ nodal spacing per temporal iteration. An approximately constant propagation speed is a characteristic of many observed epidemics in homogeneous population density environments: introduced into Italy by ship around December 1347, in the following three years the Black Death propagated steadily northward across mainland Europe by about 300 km per year.

Considering that infected agents spread the disease through their random walk on the lattice, a constant propagation speed is a curious result; the root-mean-squared displacement in a random walk increases as the square root of the number of steps taken (see appendix C.5 if in a need of a refresher on the statistics of random walks). This square-root law is plotted as a dash-dotted line in figure 9.6. It offers a reasonable representation of epidemic spread in the first 20 iterations or so (see inset), but afterward grossly underestimates the propagation speed. This is because infection does not spread through a single sick agent stumbling its way through the lattice, but rather through a sequence of successive infection events. This is like a row of toppling dominos, where the toppling wave travels much farther than any individual domino.

9.5 Epidemic Self-Organization

The *infection rate* can be measured by keeping track of how many healthy walkers are infected by each sick walker in the course of the simulation.[1] The corresponding PDFs are plotted

---

[1] In the epidemiological literature, this is called the "reproduction number."

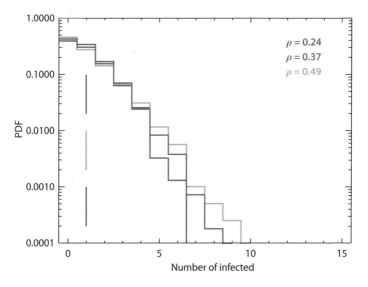

Figure 9.7. Probability density functions of infection rates for the two specific simulations of figure 9.5, having $\rho = 0.24$ and $\rho = 0.49$, and a third with an intermediate initial population density $\rho = 0.37$, as color coded. The three correspondingly color-coded vertical line segments indicate the means of the distributions, in all cases equal to unity to better than one part in 10^3. All three distributions are very much alike in shape.

in figure 9.7, for the two representative simulations plotted in figure 9.5 ($\rho = 0.24$ and 0.49), together with a third representative simulation with $\rho = 0.37$, in all three cases with $L = 20$. These are global statistics, built using all infection events irrespective of when they occurred during the epidemics.

What is truly remarkable in figure 9.7 is that, despite very different death rates, durations, and patterns of epidemic spread (as per figure 9.5), all three distributions have very similar shapes. Even more remarkable, they also all have a mean infection rate equal to unity, meaning one infected per infector, to better than one part in 10^3. Because the population density is rapidly reduced in the vicinity of epidemic fronts, locally the epidemic extinguishes itself, leading to a form of self-regulation that continuously maintains the epidemic at the edge of termination. Put differently, through local interactions (infection) and diffusive-like spreading (random walk) of infected individuals, and no matter the population density, once (and if) it gets going, the epidemic self-organizes dynamically around a marginal infection rate of exactly unity; a result both neat and unexpected, isn't it!

9.6 Small-World Networks

In all the epidemic simulations we have considered thus far, infection is a purely local process, as infected agents can infect only healthy agents located in their immediate spatial vicinity, as determined by the range of their random walk before they fall dead. This may be an appropriate first-order model for the spread of the Black Death in the Middle Ages, but is inappropriate for the spread of pandemics in our modern world. Viruses can now hitch an airplane ride and travel halfway across the world within a single day. This is why airports have become the front line of the battle against pandemics.

Pretty much everybody, out of sheer boredom, has at least once stared at the map of airline routes at the back end of that infamous airline magazine inevitably found-in-the-pocket-of-the-seat-in-front-of-you (it still makes better reading than the safety card). The airline routes are idealized as smooth curves linking one city to another, even though few planes would ever follow exactly this path. When planning a trip, connectivity often takes precedence over geographical proximity. What often matters most in choosing a ticket is how many links there are between the departure and arrival cities, as defined by the airline's *network* of connecting flights. A trip's effective "distance" is no longer measured in kilometers (or miles), but rather as the number of links required to go from one city to another. As a resident of Montréal, a hub for Air Canada, I am in fact "closer" to London (UK) than I am to London (Ontario): 5 nonstop and 62 one-stop flights per day to the former, versus 9 flights per day to the latter, all one-stop; and connections is what seasoned air travelers are most eager to keep to a minimum, because the probability of your trip (and/or luggage) going to pot because of flight delays or cancelations increases rapidly the more connections a trip involves.

Figure 9.8 depicts four possible networks linking $N = 12$ nodes. Their equidistant spacing along the perimeter of a circle is for plotting purposes only, and irrelevant for everything that follows. The network in panel A is locally connected, and a periodic equivalent of the nearest-neighbor 1-D percolation lattice introduced in section 4.1 and of the 1-D sandpile model of chapter 5. This network has 12 links, and the node-to-node distance varies from 1 to 6 links, with an average travel distance of 3.27 links for the network as a whole. The network in panel B, in contrast, is fully connected; it has $N \times (N - 1) = 132$ links, and an average travel distance of 1 link, by construction. This is the dream of any semiregular air traveler, but an airline adopting this model would face huge operating costs.

The network in panel C is a single-hub network, with all nodes connected only to a hub node, here node 0. This network has $N - 1 = 11$ links, and an average travel distance of 1.92 links.

The random-looking network in panel D has 13 links, for an average travel distance of 2.76 links. The largest average single-node travel distance is 3.5 links to/from node 2, the smallest is 1.6 to/from node 6, the longest pairwise distance is 5 links, occurring between nodes 2 and 7 and between nodes 3 and 7. The pattern of node linking is random, but not in the sense of a uniform random distribution; 5 nodes have only 1 link, 3 nodes have 2 links, 3 have 3 links, and node 6 has 6 links, acting here as a kind of hub. This network was constructed by first assigning a number of links to each node by drawing this number from a

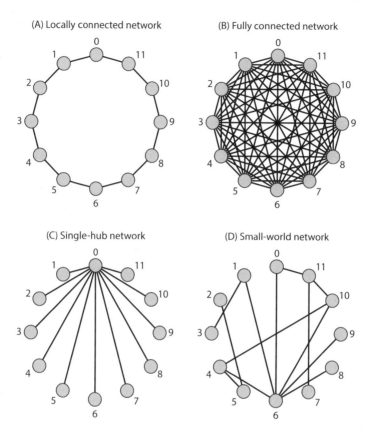

Figure 9.8. Four possible static networks connecting 12 nodes. The locally connected network in panel A introduces links only between nearest neighbors, while the fully connected network in panel B is at the opposite extreme, with every node being connected to every other node. In the single-hub network in panel C all nodes connect to each other via node 0. The random-looking "small-world" network in panel D belongs to the class of scale-free networks, with the number of links to each node distributed as a power law (see text).

power-law distribution spanning the range $[1, N]$, then picking random nodes to which to connect these links. Such a network is scale invariant, and, for reasons to be elucidated on further below, is known as a *small-world* network.

The fully connected network may have the smallest average travel distance, but if links are "expensive," the total cost is best defined as this mean travel time multiplied by the total number of links in the network. Typically, a measure of this type is what an airline would strive to minimize. For the four networks of figure 9.8, this comes out at 39.2, 132.0, 21.1, and 35.9 for panels A, B, C, and D, respectively. The single-hub network is most cost effective under this definition, with the small-world network coming in second and the local network not too far behind. Not surprisingly perhaps, the fully connected network ends up the most inefficient, by a large factor.

In the real world, connection efficiency (however defined) is not the end of the story; *redundancy* is another extremely important factor. If one link is broken or one node taken off-line for the local network of figure 9.8A, the average travel distance goes up to 4.33, but all nodes remain connected. However, break a second link, and a part of the network becomes isolated from the rest. The fully connected network in panel B is largely impervious to the loss of one or a few links: the travel time between the two disconnected nodes simply increases by 1 link, and the rest of the network remains unaffected. Typically, a large number of links must be broken before a node becomes isolated; such a network is *maximally redundant.* In the single-hub network of panel C, losing a link or the node it connects to, affects only this node, unless failure occurs at the hub; then all nodes lose connection to one another. Single-hub networks have a *single point of failure.* This is why large airlines operate more than one hub, even if it is less cost efficient than single-hub operation. When operating in single-hub mode, all it takes is one good snowstorm or bomb threat to paralyze the whole network. The small-world network in figure 9.8D is more robust in this respect. Like in the single-hub network, nodes with a single link to the network become isolated if that link breaks; but here, even in the worst-case scenario of hub node 6 going off-line, only nodes 1, 2, 8, and 9 become isolated, since nodes 4 and 10 can pick up at least some of the traffic normally going through node 6.

You can now imagine what would happen if you populated each node with a group of agents, each having a small but finite probability of traveling to another node, and infected one such agent with some horribly contagious disease (and in case you cannot, the Grand Challenge closing this chapter will lead you through it). Hub nodes would become critical, and this is where one would concentrate efforts to detect and quarantine sick individuals (and perhaps vaccinate healthy ones), in order to avoid a pandemic.

Scale-free networks pop up everywhere. The pattern of links between web pages has been argued to be a scale-free network, likewise with the pattern of citations of scientific papers, and the connectivity pattern of electrical power grids. The brain's interconnected neurons arguably make up the ultimate complex network, and it may well be scale-free (although this remains to be demonstrated). At a more mundane level, try drawing on a piece of paper the network of your own social relations, including not just a link between you and them, as in a single-hub network, but also links between mutual friends and acquaintances, as well as links to their friends and acquaintances that you are aware of, even though they are not part of your own immediate social network. Unless you live in Antarctica or are a true mountaintop hermit, pretty soon you will end up with a scale-free network of the small-world variety; and if you have on hand a very large piece of paper, you will soon realize that anyone can connect to pretty much anyone else through a surprisingly small number of links. This is why we feel that "it's such a small world" when we meet a perfect stranger and find out she happens to be good friends with the younger brother of another friend of ours living abroad; and this is also why such networks are known as "small-world" networks.

Anyhow, you may go ahead and add "catching a contagious disease" to your list of good reasons to avoid connecting flights; but also keep all of this in mind before dropping in on a friend, the next time you come down with a very nasty flu.

9.7 Exercises and Further Computational Explorations

1. The time series of infected walkers in figure 9.2 shows relatively well-defined surges of duration of order 10^2 iterations. Such repeating surges are actually often observed in real epidemics. Can you figure out what sets this characteristic timescale?

2. In discussing the dependence of epidemic characteristics on the initial population density, the resemblance between the top panel of figure 9.4 and the bottom panel of figure 4.6 was noted, suggesting that epidemic spread might be related to percolation, and thus to criticality. If this is criticality, what is the control parameter here? Could this be SOC? Would you say that the branching structure in the left panel in figure 9.5 is a fractal?

3. Use the simulation code of figure 9.1 to examine how the spread of epidemics varies with the value of the lifespan parameter L, at initial population densities $\rho = 0.25$ and $\rho = 0.5$.

4. Carry out 100 statistically distinct simulations using the parameter settings used in figures 9.2 and 9.3. Construct histograms of epidemic duration and total deaths. Are these distributions Gaussian-like? Are they even approximately symmetrical about their mean value?

5. This one is for readers with coding experience in Python—or interested in developing it. Restructure the simulation code of figure 9.1 so that it operates on *lists* of healthy and sick walkers, rather than always looping

on the whole population (including dead walkers) at every temporal iteration (see lines 25 and 38). This is most easily done by taking advantage of Python's list manipulation operators and functions, adding newly infected walkers to the "sick" list, removing from it dead walkers, etc. Think carefully how you would go about modifying the internal loops within the outer temporal loop in figure 9.1.

6. Your Grand Challenge is to simulate epidemic spread on a small-world network. You may use the network of figure 9.8D, or design your own, larger, scale-free network. On each of the network's $k = 1, \ldots, N$ nodes, place n_k agents, where n_k can be the same on all nodes (100, say), or can vary from one node to another according to your favorite recipe. Conceptually, each of these nodes is a "city," a bit like a coarse-grained version of the nearest-neighbor lattice used for the simulations presented in this chapter. At every temporal iteration, a sick agent has a probability p_i ($\ll 1$) of infecting another agent on the same node, and a probability p_t (also $\ll 1$) of traveling to a randomly chosen linked node. As in the lattice-based simulations considered in this chapter, once infected, an agent has a finite lifetime L. Introduce a single sick agent on a randomly selected node, and follow the spread of the epidemic, for various values of p_i, p_t, and initial nodal population. Once you have identified a parameter regime where the epidemic invariably takes off, fix the value of these parameters and introduce a vaccination campaign on the primary hub node. The idea is that a randomly selected fraction f of the nodal population is vaccinated, and cannot be infected or carry the disease. Determine the vaccination fraction f required to prevent epidemic spread with better than 90% probability.

9.8 Further Reading

Many historical accounts of epidemics are available in the popular literature. I have certainly not read them all, but so far my favorite remains

Zinsser, H., *Rats, Lice and History* (1935), available as a 1996 reprint, Black Dog & Leventhal Publishers.

Also well worth reading in this context is *L'oeuvre au noir* (translated into English as *The Abyss*), by Marguerite Yourcenar. Readers fluent in higher mathematics may be interested in comparing the model introduced in this chapter to the more conventional statistical and differential-equation-based approaches to the modeling of epidemic spread. Good entry points into this vast literature are

Daley, D.J., and Gani, J., *Epidemic Modeling: An Introduction*, Cambridge University Press (1999);

Murray, J.D., *Mathematical Biology*, Berlin: Springer, chaps. 19, 20 (1989).

Part 4 of Mitchell's book cited in the bibliography to chapter 1 offers an engaging and nontechnical general introduction to the science of networks. At a more technical level, see

Watts, D.J., *Small Worlds: The Dynamics of Networks between Order and Randomness*, Princeton University Press (2003);

and specifically for disease spreading on networks,

Vasquez, A., "Epidemic outbreaks on structured populations," *J. Theor. Biol.*, **245**, 125–129 (2007).

Other complex system models introduced in earlier chapters can be defined on networks; for a network-based version of the sandpile model akin to that introduced in chapter 5, see

Brummitt, C.D., Souza, R.M., and Leicht, E.A., "Suppressing cascades of load in interdependent networks," *Proc. Nat. Acad. Sci.*, **109**(12), E680–689 (2012);

these authors focus on avalanching behavior on electrical power grids described by networks of varying levels of connectivity. In a similar vein, the following presents a network-based analysis of financial system crashes:

Haldane, A.G., and May, R.M., "Systemic risk in banking ecosystems," *Nature*, **469**, 351–355 (2011).

10

FLOCKING

There is safety in numbers. Some of this is psychological, but it also harks back to basic geometry: because the surface to volume ratio of compact objects decreases with increasing size, the number of individuals exposed to predators at the edges of a large animal flock is small compared to those protected within its interior, the more so the larger the flock. Add the deterrent of perhaps looking like a dangerously large animal under suboptimal viewing conditions, and the possibility of group manoeuvres confusing an approaching predator, and you have a potential evolutionary advantage. That the advantage is real and not just potential is confirmed by the fact that a wide variety of living creatures have evolved this behavioral strategy, including many species of mammals (herds), birds (flocks), fishes (schools), and insects (swarms).

Models of flocking have also been used to understand—and control—the movement of dense human crowds in socially extreme situations. Indeed, the flocking model introduced in this chapter closely follows one developed with the specific aim of understanding global crowd movements in the so-called mosh pits at heavy metal rock concerts. Human crowd movement (and management) thus provides the context of the simulations described in what follows.

10.1 Model Definition

The flocking model considered here is defined in two spatial dimensions on the periodic unit square: $x, y \in [0, 1]$. Within this domain, N agents are moving, under the influence of four forces. The forces acting on any agent j are the following:

1. **Repulsion.** In a flock or crowd, the bodily sizes of participating individuals sets a typical lower limit to the distances between individuals (e.g., a shoulder width). To pack a crowd tighter would require a substantial external force, which would typically meet an equally substantial resistance. This property is modeled here by introducing a short-range repulsion force, which is very intense within an interaction distance r_0, but falls very rapidly at larger distances:

$$\mathbf{F}_j^{\text{rep}} = \varepsilon \sum_{k=1}^{N} \begin{cases} (1 - r_{jk}/(2r_0))^{3/2}\hat{\mathbf{r}}_{jk}, & r_{jk} \leq 2r_0, \; j \neq k, \\ 0 & \text{otherwise.} \end{cases}$$

$$(10.1)$$

   Here, r_{jk} is the distance between agents j and k, and $\hat{\mathbf{r}}_{jk}$ is a unit vector pointing from k toward j. The parameter ε sets the magnitude of this repulsion force. All simulations considered below use a spatial range $r_0 \ll 1$.

2. **Flocking.** In a crowd, many people behave like sheep, blindly following others around them. Here this ovine tendency is modeled by introducing a flocking force which tends to align the velocity vector of agent j with that of the group of individuals located within a flocking radius r_f of its own location. Mathematically,

$$\mathbf{F}_j^{\text{flock}} = \frac{\alpha \bar{\mathbf{V}}}{\sqrt{\bar{\mathbf{V}} \cdot \bar{\mathbf{V}}}}, \qquad (10.2)$$

where

$$\bar{\mathbf{V}} = \sum_{k=1}^{N} \begin{cases} \mathbf{v}_k, & r_{jk} \leq r_f, \quad j \neq k, \\ 0 & \text{otherwise}, \end{cases} \tag{10.3}$$

measures the vectorially summed velocities of all k agents located within r_f of agent j, and the parameter α sets the magnitude of this flocking force. In all simulations that follow, we use $r_f \geq 4r_0$, reflecting the fact that in a crowd we typically see (and react to) only others that are relatively close to us, these being still more numerous than those in immediate bumping range.

3. **Self-propulsion.** In some contexts, for example a protest march, some individuals are purposefully trying to move at some finite speed. This is modeled here through a self-propulsion force, defined mathematically as

$$\mathbf{F}_j^{\text{prop}} = \mu(v_0 - v_j)\hat{\mathbf{v}}_j, \tag{10.4}$$

where v_0 is the target velocity of agent j, and the parameter μ sets the magnitude of the self-propulsion force. The speed of agent j is

$$v_j = \sqrt{v_{x,j}^2 + v_{y,j}^2}, \tag{10.5}$$

and $\hat{\mathbf{v}}_j$ is a unit vector aligned with its current velocity:

$$\hat{\mathbf{v}}_{x,j} \equiv \frac{v_{x,j}}{v_j}\hat{\mathbf{x}}, \qquad \hat{\mathbf{v}}_{y,j} \equiv \frac{v_{y,j}}{v_j}\hat{\mathbf{y}}. \tag{10.6}$$

In all simulations discussed in this chapter, we set $\mu = 10$. Note that setting $v_0 = 0$ will tend to decelerate agent j, not exactly what one thinks of as "self-propulsion," but this is not unrealistic in a crowd; depending on context many people just naturally slow down to a standstill unless they are being actively pushed around.

4. **Random.** Finally, agents can be subjected to small—or not-so-small—perturbations of whatever origin. This is modeled through a randomly oriented force,

$$\mathbf{F}_j^{\text{rand}} = \boldsymbol{\eta}_j, \tag{10.7}$$

where each component of $\boldsymbol{\eta}$ is extracted from a distribution of random deviates, uniform in the range $[-\eta, \eta]$.

The total force acting on agent j is thus the vector sum of these four forces:

$$\mathbf{F}_j = \mathbf{F}_j^{\text{rep}} + \mathbf{F}_j^{\text{flock}} + \mathbf{F}_j^{\text{prop}} + \mathbf{F}_j^{\text{rand}}, \tag{10.8}$$

which will induce an acceleration according to Sir Isaac Newton's celebrated third law of motion:

$$\mathbf{a}_j = \frac{\mathbf{F}_j}{M}. \tag{10.9}$$

We can suppose that all agents have a mass $M = 1$, without loss of generality. From one time step to the next, agents move and adjust their speeds according to

$$\mathbf{x}_j(t + \Delta t) = \mathbf{x}_j(t) + \mathbf{v}_j(t)\Delta t,$$
$$\mathbf{v}_j(t + \Delta t) = \mathbf{v}_j(t) + \frac{\mathbf{F}_j(t)}{M}\Delta t, \tag{10.10}$$

with Δt the time step.[1] Most simulations reported upon in this chapter use a time step $\Delta t = 0.01$. The initial condition usually

---

[1] These expressions result directly from the application of the Euler explicit first-order finite difference formula to the differential form of Newton's laws of motion. Positional accuracy could be improved by writing

$$\mathbf{x}_j(t + \Delta t) = \mathbf{x}_j(t) + \mathbf{v}_j(t)\Delta t + \frac{1}{2}\frac{\mathbf{F}_j(t)}{M}(\Delta t)^2.$$

However, this yields an algorithm where velocities are evaluated less accurately than positions, an unwanted feature in situations where the force $\mathbf{F}$ depends not just on $\mathbf{x}$ but also on $\mathbf{v}$, which is the case here.

consists in randomly distributing a fixed number of individuals in the unit square, and assigning each a randomly oriented velocity with magnitude in some preset interval. The domain is deemed periodic in x and y, meaning, for example, that any agent moving beyond $x > 1$ immediately reappears at the left side of the domain, with the same speed (magnitude and orientation) as when leaving from the right; and similarly for agents exiting at $x < 0, y < 0$, and $y > 1$. Geometrically, it is as if the agents were moving on the surface of a torus, much like the lattice-painting ant agents encountered way back in section 2.4.

10.2 Numerical Implementation

Figure 10.1 provides a listing for a Python code that implements the above simulation algorithm. This code looks deceptively simple, as it consists of little more than initializations and implementation of equations (10.10); this is because all the action—and coding intricacies—are contained in the user-defined Python functions buffer and force, which are invoked to calculate the total force acting on every agent at every temporal iteration. Source listings of these functions are given in figures 10.2 and 10.3.[2] While positional and velocity periodicities are easy to implement (lines 40–41 in figure 10.1), the calculation of the repulsion and flocking forces near domain boundaries is where the challenge lies. This is handled as a two-step process, through the user-defined functions buffer and force.

The first step, handled by the function buffer, is to replicate agents located closer to a boundary than the range of the flocking force, so that they can effectively contribute to the flocking (and repulsion) forces felt by agents located close to the opposite

---

[2]One may note that these program subunits are as intricate as the primary code calling them. This is a common situation in real simulation codes, where one strives to define functional subunits so as to maintain a visually clear logical flow within each program unit.

```
1  # FLOCKING SIMULATION ON THE UNIT SQUARE
2  import numpy as np
3  import matplotlib.pyplot as plt
4  #-----------------------------------------------------------------------
5  N     =350                      # Number of agents
6  n_iter=1000                     # Number of temporal iterations
7  dt    =0.01                     # Time step
8  r0    =0.005                    # Range of repulsion force
9  eps   =0.1                      # Amplitude of repulsion force
10 rf    =0.1                      # Range of flocking force
11 alpha =1.0                      # Amplitude of flocking force
12 v0    =0.0                      # Target speed
13 mu    =10.                      # Amplitude of self-propulsion force
14 ramp  =0.5                      # Amplitude of random force
15 #-----------------------------------------------------------------------
16 # The buffer and force functions of Figs. 10.3 and 10.4 should go here
17 #-----------------------------------------------------------------------
18 x,y  =np.zeros(N),np.zeros(N)   # Positions of agents
19 vx,vy=np.zeros(N),np.zeros(N)   # Velocities of agents
20 fx,fy=np.zeros(N),np.zeros(N)   # Forces on agents
21 xb =np.zeros([4*N])             # Define buffer zone arrays
22 yb =np.zeros([4*N])
23 vxb=np.zeros([4*N])
24 vyb=np.zeros([4*N])
25 for j in range(0,N):            # Initialize positions and velocities
26     x[j] =np.random.uniform()   # Random position in unit square
27     y[j] =np.random.uniform()
28     vx[j]=np.random.uniform(-1.,1.) # Random velocity components in [-1,1]
29     vy[j]=np.random.uniform(-1.,1.)
30
31 for iterate in range(0,n_iter):     # Temporal loop
32     # First add replicate agents in buffer around unit square
33     nb,xb,yb,vxb,vyb=buffer(max(r0,rf),x,y,vx,vy)
34     # Now calculate acceleration for each real agent
35     fx,fy=force(nb,xb,yb,vxb,vyb,x,y,vx,vy)
36     vx+=fx*dt                    # Eqs (10.10) by components
37     vy+=fy*dt                    # (remember mass=1)
38     x +=vx*dt
39     y +=vy*dt
40     x =(1.+x) % 1                # Enforce periodicity in x
41     y =(1.+y) % 1                # Enforce periodicity in y
42 # End of temporal loop
43 plt.scatter(x,y)                 # Plot positions of agents
44 plt.quiver(x,y,vx,vy,headlength=5)  # Plot velocity vectors of agents
45 plt.axis([0.,1.,0.,1.])
46 plt.show()
47 # END
```

Figure 10.1. Basic Python code for the flocking model used in this chapter. This code requires two user-defined functions called buffer and force, as listed in figures 10.2 and 10.3.

```
1   # BUFFER FUNCTION: INTRODUCE REPLICATE AGENTS OUTSIDE OF UNIT SQUARE,
2   # IN A SQUARE BUFFER EXTENDING OUTWARD BY A DISTANCE rb
3   def buffer(rb,x,y,vx,vy):
4       xb[0:N],yb[0:N]   =x[0:N],y[0:N]              # Initialize buffer arrays
5       vxb[0:N],vyb[0:N]=vx[0:N],vy[0:N]
6       nb=N-1                                        # Already have N real agents
7       for k in range(0,N):                          # Add replicants to buffer
8           if ( x[k] <= rb ):                        # Close to left
9               nb+=1
10              xb[nb]=x[k]+1.
11              yb[nb],vxb[nb],vyb[nb]=y[k],vx[k],vy[k]
12          if ( x[k] >= 1.-rb):                      # Close to right
13              nb+=1
14              xb[nb]=x[k]-1.
15              yb[nb],vxb[nb],vyb[nb]=y[k],vx[k],vy[k]
16          if ( y[k] <= rb ):                        # Close to bottom
17              nb+=1
18              yb[nb]=y[k]+1.
19              xb[nb],vxb[nb],vyb[nb]=x[k],vx[k],vy[k]
20          if ( y[k] >= 1.-rb ):                     # Close to top
21              nb+= 1
22              yb[nb]=y[k]-1.
23              xb[nb],vxb[nb],vyb[nb]=x[k],vx[k],vy[k]
24          if ( x[k] <= rb and y[k] <= rb ):         # Close to bottom left
25              nb+= 1
26              xb[nb],yb[nb]=x[k]+1.,y[k]+1.
27              vxb[nb],vyb[nb]=vx[k],vy[k]
28          if ( x[k] >= 1.-rb and y[k] <= rb ):      # Close to bottom right
29              nb+= 1
30              xb[nb],yb[nb]=x[k]-1.,y[k]+1.
31              vxb[nb],vyb[nb]=vx[k],vy[k]
32          if ( x[k] <= rb and y[k] >= 1.-rb ):      # Close to top left
33              nb+= 1
34              xb[nb],yb[nb]=x[k]+1.,y[k]-1.
35              vxb[nb],vyb[nb]=vx[k],vy[k]
36          if ( x[k] >= 1.-rb and y[k] >= 1.-rb ):   # Close to top right
37              nb+= 1
38              xb[nb],yb[nb]=x[k]-1.,y[k]-1.
39              vxb[nb],vyb[nb]=vx[k],vy[k]
40      # End of buffer loop
41      return nb,xb,yb,vxb,vyb                       # Total real+replicate agents
42  # END OF FUNCTION BUFFER
```

Figure 10.2. A used-defined Python function used in the flocking model to add agent replicates outside the unit square domain, to facilitate the computation of the total force carried out by the user-defined function force listed in figure 10.3. This function is called once at the beginning of each temporal iteration (line 33 in figure 10.1).

```
1   # FORCE FUNCTION: CALCULATE TOTAL FORCE ACTING ON ALL AGENTS
2   # Values for r0,eps,rf,alpha,v0,mu,ramp set in calling program
3   def force(nb,xb,yb,vxb,vyb,x,y,vx,vy):
4
5       for j in range(0,N):                     # Loop over real agents
6           repx,repy,flockx,flocky,nflock=0.,0.,0.,0.,0
7           for k in range(0,nb):                # Loop over agents+replicants
8               d2=(xb[k]-x[j])**2+(yb[k]-y[j])**2 # Squared distance j,k
9               if (d2 <= rf**2) and (j != k):   # k contributes to flocking
10                  flockx+=vxb[k]
11                  flocky+=vyb[k]
12                  nflock+=1
13              if (d2 <= 4.*r0**2):             # k contributes to repulsion
14                  d=np.sqrt(d2) ;              # Distance between j and k
15                  repx+=eps*(1.-d/(2.*r0))**1.5 *(x[j]-xb[k])/d # Eq (10.1)
16                  repy+=eps*(1.-d/(2.*r0))**1.5 *(y[j]-yb[k])/d
17          # End of loop over agents and replicants
18
19          normflock=np.sqrt(flockx**2+flocky**2) # Denominator in Eq (10.2)
20          if ( nflock == 0 ): normflock=1.    # To avoid 0/0 division
21          flockx=alpha*flockx/normflock       # Flocking Eq (10.2)
22          flocky=alpha*flocky/normflock
23          vnorm =np.sqrt(vx[j]**2+vy[j]**2)    # Speed of agent j
24          fpropx=mu*(v0-vnorm)*(vx[j]/vnorm)   # Self-propulsion Eq (10.4)
25          fpropy=mu*(v0-vnorm)*(vy[j]/vnorm)
26          frandx=ramp*np.random.uniform(-1.,1.) # Random force Eq (10.5)
27          frandy=ramp*np.random.uniform(-1.,1.)
28          fx[j]=(flockx+frandx+fpropx+repx)    # Total force on agent j
29          fy[j]=(flocky+frandy+fpropy+repy)
30      # End of loop over real agents
31
32      return fx,fy
33  #END OF FORCE FUNCTION
```

Figure 10.3. A user-defined Python function that computes the total force acting on each agent (j-indexed loop starting on line 5). This function is called at every temporal iteration, at line 35 in the temporal loop in the flocking simulation code listed in figure 10.1.

boundary. The idea is illustrated in figure 10.4. Here, 16 agents (solid black dots) populate the unit square (in black). Now define a buffer area (gray shading) corresponding to the periodic unit domain with its x and y boundaries expanded outward by a distance equal to the flocking radius r_f. Particles within the unit square, but closer than r_f to a boundary, get replicated one unit away inside this buffer, in the direction opposite to that of the nearby boundary. This replication process is indicated by the color coding of replicated agents in figure 10.4. Note how agents

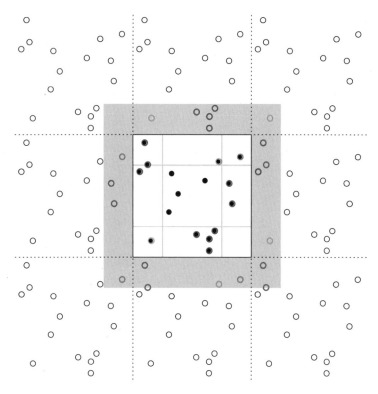

Figure 10.4. Construction of a buffer zone (gray shading) with replicated agents (colored open circles), to ensure proper calculation of the flocking and repulsion forces experienced by agents (solid black dots) distributed over a periodic unit domain (black square). Any agent located closer than the flocking radius from a boundary (solid dots with colored ring) gets replicated a unit distance away in the opposite direction, as captured here by the color coding. Note how agents located close to a corner get replicated thrice. For the specific distribution of 16 agents in the unit square shown here, a total of 20 replicates have been created.

located close to a corner of the unit square spawn three replicates: horizontally, vertically, and diagonally. Here the computation of the flocking (and repulsion) forces acting on any one of the

16 original agents located within the unit squares could now involve up to $15 + 20$ other agents, real or replicates.

Examine carefully the two functions listed in figures 10.2 and 10.3 and note the following:

1. The job carried out by function buffer is to construct expanded arrays xb, yb, vxb, vyb for the positions and velocities of agents and replicates, as per figure 10.4 and the accompanying discussion. The buffer width rb is passed through the function's argument list when invoked by the main program (line 33 in figure 10.1), where it is set to the flocking radius r_f.

2. The (modified) position and velocity of every such replicated agent are introduced at position nb of the above expanded arrays, with nb incremented by 1 every time a replicated agent is added. At the end of this operation, the arrays xb, yb, vxb, vyb contain nb ($>$ N) agents, distributed in the interval $[-r_b, 1 + r_b]$ in x and y.

3. The force acting on agent j is calculated based on its distance to every other agent, including replicates. This means that the j-indexed loop, starting at line 5 in the force function of figure 10.3, runs from 0 to $N - 1$, i.e., only over the N "real" agents located within the unit square. The k-indexed loop within force (lines 7–16 in figure 10.3), however, runs from 0 to nb $- 1$.

4. There are now N $\times$ (nb $- 1$) pairs of (distinct) agents between whom distance-based forces must be calculated, at every temporal iteration. This calculation better be as efficient as possible. A first test (line 9) checks whether agent k is within the flocking radius r_f of agent j; if so, the flocking force is calculated, and then a second test (line 13) verifies whether k is also within $2r_0$ of j, in which case its contribution to the

repulsion force is also calculated. A consequence of this construct is that the first `if` will be executed nb times per real agent, but the second only a few times, since typically only a few agents are within a radius r_f of agent j.

5. The calculation of the flocking and repulsion forces includes a test (in line 9) that prevents computing the repulsion force of an agent on himself should $j = k$. Look again at equation (10.1) and imagine what would happen without this exclusion.

6. If no agent is within the flocking radius r_f of agent j, then the calculation of the flocking force will produce a division by zero, since we then have the norm $\sqrt{\bar{\mathbf{V}} \cdot \bar{\mathbf{V}}} = 0$ in equation (10.2); to avoid this problem the counter variable `nflock` tallies up the number of agents within r_f of agent j (line 12); if this is zero, then the norm (local variable `normflock`) is artificially set to unity (line 20), so that the flocking force will be zero, rather than whatever you get from dividing zero by zero (with many computing languages you would get that (in)famous `NaN`).

7. The x- and y-components of the flocking and repulsion forces are calculated separately and accumulated in the local variables `flockx`, `flocky`, `repx`, and `repy`. It is only upon exiting the k-indexed inner loop that the total forces are calculated, including the contributions of the purely local self-propulsion (lines 24–25) and random (lines 26–27) forces.

8. Upon returning control to the calling program unit, the final step consists in using equation (10.10) to update the position and velocity arrays for all agents j (lines 36–39 in figure 10.1), with periodicity enforced (lines 40–41), and without forgetting that we have assumed all agents to have unit mass ($M = 1$ in

equation (10.9)). Note here the use of mathematical operators acting on NumPy arrays, rather than array elements within a loop.

Even with the little tricks introduced here, such brute-force computing of distances between all pairs of agents can become prohibitively expensive as N gets very large. There exist algorithms far more efficient for this, developed for so-called N-body simulations. The interested reader will find a few good entry-point references at the end of this chapter.

10.3 A Behavioral Zoo

With four forces acting in the simulations and the large number of numerical parameters defining their respective ranges and magnitudes, it is no surprise that the model can produce a very wide range of global behaviors. For convenience and later reference, table 10.1 lists all model parameters and the corresponding numerical values used in the various sets of simulations presented in the remainder of this chapter.

Rather than taking our customary detailed look at one specific simulation, in the present context it will prove more useful to first consider a few simple simulations demonstrating the action of a subset of forces, to better appreciate the behavior of subsequent simulations.

Figure 10.5 shows snapshots of four simulations with the flocking force turned off, and self-propulsion acting to brake the individuals to rest (target speed $v_0 = 0$ in equation (10.4)); numerical values for other model parameters are given in the caption and listed in the fourth column of table 10.1. What varies in this sequence is the magnitude of the random force, increasing from left to right, as labeled. The top row shows the positions of all agents after an elapsed time interval $t = 20$, from an initial condition consisting of 100 agents randomly distributed in the unit square, with randomly oriented initial velocities.

Table 10.1. Model parameters (active/passive where appropriate)

	Description	Eq.	10.5	10.6	10.7	10.8
					Figure	
r_0	repulsion radius	(10.1)	0.05	—	0.025	0.025
ε	repulsion amplitude	(10.1)	25	0	25	25
r_f	flocking radius	(10.3)	—	0.1	0.1	0.1
α	flocking amplitude	(10.2)	0	1	1	0.1/1
v_0	target velocity	(10.4)	0	0	0.02/0	0.05/0.02
μ	self-propulsion amplitude	(10.4)	10	10	10	10
η	random force amplitude	(10.7)	1,3,10,30	0.1	0.1/0	10/0.1
N	number of agents		100	342	114–456	342

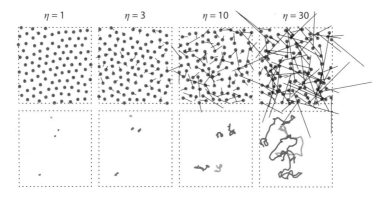

Figure 10.5. Four simulations driven only by repulsion and random forces, with self-propulsion acting as a brake ($v_0 = 0$). All simulations have $r_0 = 0.05$, $\varepsilon = 25$, $\alpha = 0$, $\mu = 10$, and a random force of amplitude η increasing from left to right, as labeled. The top row shows the distribution of agents, with the line segments indicating the orientations and magnitudes of their velocity vectors. The bottom panels show the trajectories of four selected agents over a time interval $100\Delta t = 2$ in the course of the simulation.

As long as the random force remains relatively small ($\eta = 1$; left column), the repulsion force rapidly pushes the agents into a quasi-equilibrium, a geometrically ordered configuration in which the total repulsion force on any agent vanishes. Because the repulsion force is isotropic, the resulting global end state must be also, which has led to a close-packed, hexagonal pattern. Here, this pattern includes some "defects" and "holes," because a few additional agents would be needed here to construct a truly regular periodic hexagonal "crystal." At low η the individual agents also move about their equilibrium position under the action of the random force, damped by the action of the self-propulsion force which acts here as a brake ($v_0 = 0$). As the trajectories plotted in the bottom row show, this motion is hardly discernable at $\eta = 1$, but already at $\eta = 3$ it is sufficient to produce noticeable perturbations in the hexagonal configuration.

At $\eta = 10$ the random force is large enough for pairs of "colliding" agents to occasionally exchange positions, leading to slow, irregular, pseudo-random drift across the domain. At $\eta = 30$ the simulation is now in a "fluid" phase, with agents describing what, to all intents and purposes, is a 2-D random walk.

In general, the spatial density of agents distributed over the domain is a key parameter in these types of simulations. In view of the short range and high intensity of the repulsion force (see equation (10.1)), one can consider that each agent bodily occupies a "surface" $\simeq \pi r_0^2$. The *compactness coefficient* (C) is defined as the ratio of the total surface collectively occupied by agents to the available surface. Since the simulation is defined on a unit square, we have

$$C = \pi N r_0^2. \tag{10.11}$$

For the $N = 100$ simulations of figure 10.5, with $r_0 = 0.05$, this gives $C = 0.785$, confirming the visual impression that agents are pretty tightly packed.

Figure 10.6 shows a simulation now driven only by the flocking force, with self-propulsion acting as a brake ($v_0 = 0$). Repulsion is turned off ($\varepsilon = 0$), and a small random force is included ($\eta = 0.1$). The first snapshot, taken at $t = 0.5$, shows how the initial random velocities are rapidly damped by the braking force, but with the flocking force already starting to align the velocity vectors of neighboring agents. By $t = 1.0$, the flocking force has led to a general acceleration of most agents, with groups of agents merging to produce a clockwise vortex on the left, which persists until about $t = 2$, by which time agents are moving as a long sinuous stream. The periodic boundary conditions lead to a "collision" at $t = 6$ as the upward-moving front of the stream on the right merges with its middle part, leaving the domain diagonally through the bottom-left corner to reappear at the upper right. This causes a merging of the stream into a single, denser flock, which ends up moving at constant

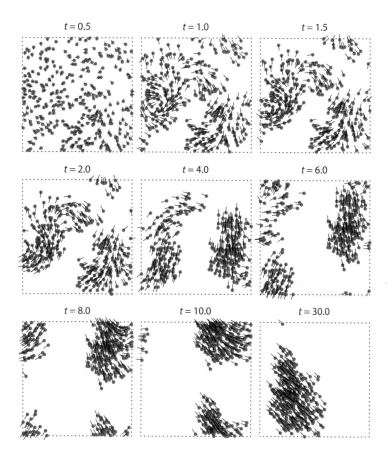

Figure 10.6. Flock formation in a simulation driven only by the flocking force, with the self-propulsion force acting as a brake ($v_0 = 0$). The parameter values for these simulations are listed in the fifth column of table 10.1. Note that the various frames are not equally spaced in time. Also keep in mind that the simulation domain is periodic in x and y (horizontally and vertically).

speed along a straight line pointing approximately northwest here ($t = 30$). This final streaming direction is ultimately determined by the initial conditions, with all directions being, in principle, equiprobable.

10.4 Segregation of Active and Passive Flockers

The variety of behaviors that can be generated in our flocking model becomes even larger if we allow for the coexistence of agents following distinct sets of dynamical rules; think, for example, of a bunch of riot-control law enforcers moving into a crowd of protest marchers; or of a group of tardy concert goers trying to push their way to the front of the general admittance floor. In such a situation we can identify "active" agents, trying to do something, and "passive" agents, not doing much until they get pushed around or hit on the head. Such a dual population of agents is readily accommodated within the simulation code of figures 10.1 and 10.3, by introducing suitable arrays of length N for the model parameters, which have different values for the two types of agents.

An interesting and important issue in crowd management is to understand under which circumstances two intermingled populations of active and passive agents can spontaneously segregate, by regrouping into distinct flocks. Figure 10.7 shows snapshots taken far into a set of four simulations, in all cases including the same number $N_a = 45$ active agents, in red, and an increasing number of their passive cousins, in green. Except for the numbers of passive agents, all simulations use the same parameter values, as listed in table 10.1 for figure 10.7. Here, active agents differ only in having a finite target velocity $v_0 = 0.02$ and being subjected to a small random force $\eta = 0.1$.

At low compactness ($C \leq 0.25$), the self-propelling active agents flock into a long stream that clears a path through the passive agents, most of the latter remaining at rest unless they happen to be pushed around by an active agent. At intermediate compactness ($C \simeq 0.5$), sustained flocking turns out to be difficult, as small flocks of active agents continually merge and separate again as they encounter channels between passive groups. Once compactness reaches two-thirds (for this parameter regime),

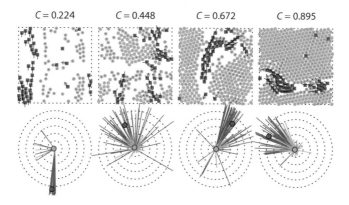

Figure 10.7. Flock formation in a sequence of simulations with compactness increasing from left to right, as labeled. The top row of panels shows the spatial distributions of active (red) and passive (green) agents after an elapsed time of 50 time units. The bottom panels show the corresponding polar diagrams of agent velocities, measured in units of the active agents' target speed v_0, and with the dotted circles indicating multiples of v_0 in steps of unity. The large colored dots indicate the mean speed of active and passive agents. The parameter values for these simulations are listed in the sixth column of table 10.1.

the groups of motionless, passive agents are sufficiently dense and massive to strongly resist entry by a self-propelled active agent, which ends up again favoring the formation of a large flock of active agents collectively succeeding in opening a channel through the crowd. The system behaves here like a two-phase flow, with the active agents percolating through a largely inert irregular matrix. At even higher compactness (rightmost panel), isolated active agents can become trapped in the close-packed "crystalline" assemblage of passive agents.

The global behavior, namely, the capacity of the active agents to flock, clearly shows a nontrivial relationship to compactness, as a consequence of the dynamical rules governing the inter- actions. This can be appreciated upon examining the velocity distributions of all agents, plotted in the bottom row of figure 10.7 in the form of polar plots, where each color-coded line corresponds to the velocity vector of one agent. At low compactness the velocities of active agents are strongly coaligned, which provides a sustained flocking force maintaining the motion. Significant scatter is present at high compactness, mostly due to agents at the edges of the flock being deflected by collisions with the solid walls of passive agents on either side of the open channel cut by the flock of active agents. The largest scatter is found at intermediate compactness, a consequence of the fact that active agents fail to form a persistent large flock.

10.5 Why You Should Never Panic

Imagine this: It's a nice Sunday afternoon and your favorite home team is facing the archenemy from elsewhere for a spot in the semifinals, so the stadium is packed solid. About halfway into the game a fire breaks out; or an earthquake suddenly starts rattling hard; or the PA system turns on to page Agent Smith to go meet the quarantine team at entrance A-8, quickly and without touching anyone or anything please, because your Ebola test turned out positive; or whatever. At any rate, such events are more likely than not to trigger a mass movement toward the stadium's exits. We all know the drill: stay calm, walk fast but don't run, no pushing, and do not use the elevators. However, based perhaps on experience—and if not, at least on what we learned from our examination of traffic flow in chapter 7—we also know that a few panicked bozos running around randomly and bumping into people can seriously disrupt what would otherwise be an orderly evacuation.

Our flocking model is ideally suited to investigating the perturbing effects of panicked individuals on collective, ordered motion. We consider again two types of agents: (1) strongly flocking ($\alpha = 1$) "calm" agents, subjected to self-propulsion to a moderate "walking" speed ($v_0 = 0.02$) and small random force

($\eta = 0.1$), and (2) "panicked" agents striving for running speed ($v_0 = 0.05$), undergoing sudden and erratic changes in direction, modeled here through a large random force ($\eta = 10$), and far less interested in flocking ($\alpha = 0.1$). All other parameter values are as listed in the rightmost column of table 10.1. The idea is thus to carry out simulations at relatively high compactness, $C = 0.67$, varying the proportion $f = N_p/N$ of panicked agents in the population, with this ratio f remaining fairly small.

Figure 10.8 shows a sequence of simulations where the fraction of panicked agents increases from zero (on the left) to a mere 5% (on the right). In the absence of panicked agents, a generally constant cruising speed is reached, with the self-propulsion force equilibrating the flocking force. The small dispersion in the orientation of velocity vectors again reflects the action of the weak random force, and the intermittent action of the repulsion force resulting from inhomogeneities in the spatial distribution of the moving flock of agents. As one would have expected, this dispersion gradually increases as more and more panicked agents are introduced into the simulation. Notice how panicked agents tend to carve out "holes" for themselves within the moving flock of calm agents, a phenomenon observed in real crowds. This is due to the repeated collisions with surrounding calm agents, driven by the random force and mediated by the repulsion force.

Probably not expected at all is the fact that even a few percent of panicked agents can induce long-term, global changes in the moving flock, and more specifically, significant changes in the spatial orientation of its motion. This is further illustrated in figure 10.9, showing the trajectories of a single calm agent in each of the four simulations of figure 10.8, plus two other trajectories at a higher fraction of panicked agents, as labeled. Even at the highest panicked fraction, these trajectories remain representative of the moving flock as a whole. It is remarkable that even as little as 2% of panicked agents can cause a deflection of the moving flock by almost 45 degrees. Of course, different deflections would

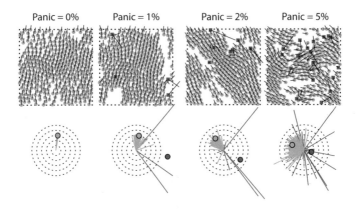

Figure 10.8. Similar in format to figure 10.7, but this time for a sequence of $C = 0.67$ simulations with an increasing fraction of panicked agents (in red). The snapshots are taken at time $t = 50$, and all simulations again use the exact same initial conditions for the positions and velocities of all agents, irrespective of their panicked or calm status.

be produced if different random initializations were used, but the trends observed in figure 10.9 are robust: flock deflection increases rapidly with increasing fractions of panicked agents, and sets in very early in the simulation. At the highest fractions of panicked agents, the net distance traveled also decreases markedly, which is not a good thing if rapid evacuation of the crowd is hoped for.

Written in big bright letters on the back of the authoritative *Hitchhiker's Guide to the Galaxy* is the well-known first rule of galactic survival: DON'T PANIC. Our flocking simulations demonstrate that this dictum also bears following even in more Earthly stressful circumstances. More importantly perhaps, the self-organization of an initially randomly moving population of agents into coherently moving flocks represents another instance of an emergent phenomenon encountered before: order from disorder through local nonlinear dynamics.

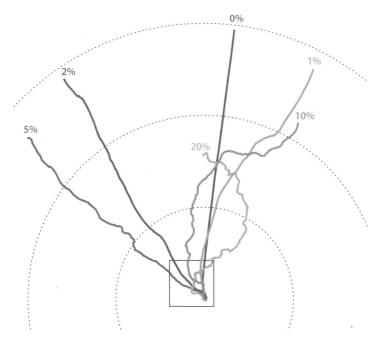

Figure 10.9. Trajectories of a randomly selected calm agent in the simulations of figure 10.8, augmented by two simulations using panicked fractions of 10% and 20%, as labeled. The dotted circles are drawn at radii of 2, 4, and 6, centered on the initial positions of the selected agents. For plotting purposes, the unit square has been replicated so as to show the trajectories in "physical" space. All trajectories cover the same time span, namely, 50 time units.

10.6 Exercises and Further Computational Explorations

1. The force function of figure 10.3 could run twice as fast by taking into account the fact that the repulsion and flocking forces of agent k on agent j are equal in magnitude but opposite in direction to the repulsion and flocking forces of agent j on agent k—as per Newton's famous action–reaction principle. Give it a go!

2. Carry out a sequence of simulations like those plotted in figure 10.5. For each, compute the final total kinetic energy, namely, the sum of $\frac{1}{2}Mv^2$ over the whole population at your last time step, and examine how this varies with η. Is the transition from "solid" to "fluid" taking place abruptly or gradually? Could this "phase transition" be considered an instance of a critical phenomenon?

3. Construct a new set of simulations such as in figure 10.7, but gradually decrease the amplitude of the flocking force (parameter α) for the active agents. At which value of α do you cease to form flocks? Is the transition abrupt or gradual? Does it depend sensitively on compactness?

4. The formation of long-lasting coherent structures, such as the (transient) vortex of figure 10.6, also takes place in two-population versions of the model. Try to look for such structures in simulations at high compactness ($0.9 \leq C \leq 1.0$), and a proportion of active agents $N_a/N = 1/3$. Active agents have small but finite target velocities ($v_0 = 0.02$) and random force ($\eta = 0.1$), while passive agents have $v_0 = 0$ and $\eta = 0$. You may vary the magnitude of the flocking force (parameter α) and self-propulsion amplitude (μ) for active and passive agents. For the other model parameters, use the values listed in table 10.1 for figure 10.7.

5. Another important task in crowd control is how to intervene to get a large compact crowd of passive or disoriented individuals to start moving collectively in a specific direction. The idea is basically the same as in figure 10.7, namely, to introduce a population of self-propelled active agents in a dense group of passive agents. Modify the self-propulsion force so that the target speed of active agents is oriented in the positive y-direction (say), and use the difference in

the average y-component of the velocity of the passive and active agents as a measure of "coupling." Identify in which portion of the model's parameter space this coupling is the strongest. Use the same parameter values as in the simulations of figure 10.7, but explore the effects of varying v_0, α, μ, and/or η for the active agents. How sensitive are your results to compactness?

6. The Grand Challenge for this chapter is a real fun one: repeat the simulation of figure 10.6, but now add a single, rapidly moving ($v_0 = 0.5$) strongly flocking ($\alpha = 5$) "predator" agent that generates a *long-range* repulsive force ($r_0 = 0.1$, say) in the flocking "prey" agents. Give the predator (and only the predator) a flocking radius 50% larger than its repulsion radius, so it can "see" and track the prey before scaring it away. Adding a moderate random force ($\eta = 1$) to the predator yields nicer results. You should observe flock shapes and evolution-resembling observations, including arched thinning flocks dividing to "confuse" the predator.

10.7 Further Reading

The flocking model introduced in this chapter is taken from

Silverberg, J.L., Bierbaum, M., Sethna, J.P., and Cohen, I., "Collective motion of humans in mosh and circle pits at heavy metal concerts," *Phys. Rev. Lett.*, **110**, 228701 (2013).

The following is also very interesting on the broader topic of crowd behavior and management:

Moussaid, M., Helbing, D., and Theraulaz, G., "How simple rules determine pedestrian behavior and crowd disasters," *Proc. Nat. Acad. Sci.*, **108**, 6884–6888 (2011).

There exists a vast biological and ecological literature on flocking; at the nonmathematical level I much enjoyed

Partridge, B.L., "The structure and function of fish schools," *Scientific American*, **246**(6), 114–122 (1982);

Feder, Toni, "Statistical physics is for the birds," *Physics Today*, **60**(10), 28 (2007).

On algorithms for N-body simulations I found the following very informative, even though it focuses on gravitational problems:

Trenti, M., and Hut, P., "N-body simulations (gravitational)," *Scholarpedia*, **3**(5), 3930 (2008).

This is available online, open access (March 2015):

http://www.scholarpedia.org/article/N-body_simulations_ (gravitational).

11

PATTERN FORMATION

11.1 Excitable Systems

Many physical, chemical, and biological systems can be categorized as *excitable*; in the simplest such systems, two "components" interact in such a way as to alter each other's state, through (nonlinear) processes of inhibition or amplification. Starting from a homogeneous rest state, many systems of this type can spontaneously generate persistent spatiotemporal patterns when subjected to some perturbation. Examples abound in chemistry, notably with autocatalytic chemical reactions. Consider the following generic chemical reaction chain taking place in a fully mixed environment:

$$A \rightarrow X, \tag{11.1}$$
$$B + X \rightarrow Y + D, \tag{11.2}$$
$$2X + Y \rightarrow 3X, \tag{11.3}$$
$$X \rightarrow C. \tag{11.4}$$

The first reaction produces reactant X by dissociation of some compound A available in large quantities; A thus provides a constant-rate source of X. The second reaction produces a second reactant Y from X through a reaction involving a compound B

also available in large quantities. The third reaction is the critical one; it converts Y back to X through a three-body reaction involving two X; the overall rate is therefore proportional to the *square* of the concentration of X in the mixture, times the concentration of Y. This one is the autocatalytic reaction in the chain: X reacts with itself to produce more of itself. The fourth reaction represents the "spontaneous" dissociation of X at some fixed rate, and acts as a sink of X. The chain as a whole converts A and B to C and D, with X and Y being produced and destroyed as intermediate steps in the chain.

If the concentrations of A and B are held fixed in the mixture (e.g., by continuous replenishment and stirring), it can be shown that there exists an *equilibrium state* where the concentrations of X and Y also remain fixed, at values

$$X_{eq} = A, \qquad Y_{eq} = B/A, \tag{11.5}$$

assuming all four above reactions have the same time constants and with reverse reaction rates set to zero. In this equilibrium state, the second reaction produces Y at the same rate as the third one destroys it, so that the concentration of X stabilizes at a level such that the chain as a whole simply converts A and B to C and D at a constant rate. However, for some values of A and B this equilibrium state turns out to be unstable, and this is due to the nonlinearity characterizing the third, autocatalytic reaction in the chain. Because reaction (11.3) proceeds at a rate proportional to X^2, while the second reaction is instead linearly proportional to X, an increase of whatever origin in the concentration of X will favor the third reaction over the second. The concentration of X will thus keep increasing, leading to runaway production of X; this runaway cannot go on forever because it also depletes Y, at a rate higher than the second reaction can replenish it. As Y plummets to a low concentration, reaction (11.3) turns off, and Y starts rebuilding through the second reaction, a reaction now favored by the high concentration of X in the mixture. This

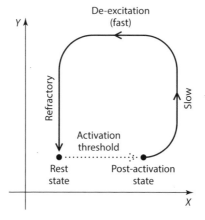

Figure 11.1. Schematic representation of an activation–recovery cycle in the [X, Y] phase space of a generic excitable system. Here, X and Y represent the excitation and recovery variables, respectively. The rest state is stable with respect to small perturbations in either X or Y, but a large perturbation exceeding the activation threshold for the variable X (dotted arrow) can initiate a large excursion in phase space, which represents the only dynamically allowed path from the postactivation state back to the rest state.

leads to a *chemical oscillation*, whereby the concentrations of X and Y wax and wane periodically. Such chemical oscillations are observed in the laboratory, the Belousov–Zhabotinsky reaction being the classical example. In that case, the excitation variable X is the concentration of bromic acid, and the recovery variable Y is the concentration of some suitable metallic ion, such as ferroin.

In the above reaction chain, X acts as an *activation variable*, and Y as a *recovery variable*, and systems capable of producing such nonlinear oscillations (or runaway) when perturbed away from their equilibrium state are deemed *excitable*. Their dynamical behavior becomes particularly interesting when excitation involves a threshold.

Figure 11.1 illustrates schematically the workings of an activation–recovery cycle in a generic excitable system, with X as the excitation variable and Y the recovery variable. The system possesses a stable *rest state*, stable in the sense that small perturbations in either X or Y are damped so that the system remains in this rest state. However, a sufficiently large perturbation in the excitation variable X (dotted arrow) can push the system into a postactivation state characterized by a value of X that now allows the growth of Y. The growth of the recovery variable Y is however restricted to a finite range, and when the upper end of this range is attained (saturation), de-excitation takes place. This drives X back to a value at which Y can no longer grow. Typically, during this *refractory stage* the system cannot be excited, and both X and Y simply return to the rest state.

In many situations of interest the activation variable reacts to system changes on a much faster timescale than does the recovery variable; the former can thus be assumed to take on one of two possible states, active or inactive, and the period of the activation–recovery cycle becomes set by the reaction timescale for the recovery variable Y. In other words, in figure 11.1 the horizontal displacements are "fast," while vertical displacements are "slow."[1]

Quite obviously, triggering the activation–recovery cycle requires some mechanism to push the excitation variable X

---

[1] Figure 11.1 is a simplified, schematic representation of the phase space plot for a system of two coupled nonlinear differential equations of the generic type

$$\frac{\partial X}{\partial t} = f(X, Y), \qquad \frac{\partial Y}{\partial t} = g(X, Y);$$

in the case of the reaction chain considered above, $f(X, Y) = A - (B + 1) X + X^2 Y$ and $g(X, Y) = BX - X^2 Y$. The rest state corresponds to the intersection of the two nullclines $f(X, Y) = 0$, $g(X, Y) = 0$. In situations where the activation variable X reacts rapidly and remains in quasi-equilibrium $(dX/dt \simeq 0)$, the phase space path for the inhibition–recovery cycle follows the $f = 0$ nullcline in response to (slow) variations in the recovery variable Y. If the $f = 0$ nullcline is multivalued in X for some range of Y, then the system can also "jump" horizontally from one branch of the nullcline to another, resulting in the type of activation–recovery cycle illustrated in figure 11.1.

beyond its activation threshold. This mechanism can certainly be external to the system. In the context of autocatalytic chemical reactions, for example, this could be as simple as an Agent Smith pouring more chemicals into the test tube. A far more interesting situation is one in which the system is spatially extended and characterized by chemical concentration gradients. Diffusion can then move chemicals from regions of higher concentration to neighboring regions of lower concentration, and in doing so activate the system away from the rest state in spatially localized regions of the domain.[2]

Dynamically similar activation–recovery cycles have been observed in contexts other than chemical reactions. A particularly interesting example is provided by electrically excitable biological tissues, such as the heart muscle or nerve axons, for which membrane potential acts as the excitation variable X, and cross-membrane ionic currents define the recovery variable Y.

11.2 The Hodgepodge Machine

The mathematical investigation of pattern formation in reaction–diffusion systems was initiated by Alan Turing, during the final years of his tragically short life. Not only was Turing an outstanding mathematician, but in the late 1940s and early 1950s he also had access to one of the earliest working computers, operating at the University of Manchester in England. He used this opportunity to generate numerical solutions of coupled nonlinear reaction–diffusion partial differential equations, at the time a complete terra incognita, since such systems are largely impervious to conventional pencil-and-paper mathematical techniques. Turing could show that reaction–diffusion systems

---

[2]In such a case, the temporal evolution of X and Y can be described mathematically by a pair of coupled partial differential equations, with the coupling nonlinearity as in footnote 1, and linear diffusion terms for X and Y, typically of the usual Fickian variety ($\propto \nabla^2 X$ and $\nabla^2 Y$).

can spontaneously generate spatial patterns, which he dubbed "chemical waves." In 1952, and once again well ahead of his time, he proposed that such chemically driven spatial pattern formation represented a key mechanism for morphogenesis in the developing embryo.

Even with the staggering increase in computing power having taken place since Turing's pioneering investigations, the mathematical and numerical investigation of spatially extended nonlinear reaction–diffusion equations remains a very computationally demanding endeavor. The *hodgepodge machine* is a simple CA that captures much of the pattern-forming behavior of the class of coupled systems of nonlinear reaction–diffusion partial differential equations of the type Turing investigated, as well as of other excitable systems in the broader sense.

The model is defined over a 2-D regular Cartesian lattice, with 8-nearest-neighbor connectivity (top+down+right+left+diagonals). The state variable s, representing the concentration of a chemical reactant, is defined as a positive integer quantity restricted to the range $0 \leq s \leq A$, where A is the *activation threshold*. A nodal value $s = 0$ corresponds to the *rest state*, $s = A$ is the *active state*, and integer values in between represent *recovery states*. Denoting by $s_{i,j}^{n}$ the state value of node (i, j) at temporal iteration n, we first define the quantities

- N_a: the number of neighboring nodes that are in the active state ($s_{i,j}^{n} = A$) at the current iteration;
- N_r: the number of neighboring nodes that are in recovery states ($0 < s_{i,j}^{n} < A$) at the current iteration;
- S: the sum of nodal values over all neighboring nodes, including node $s_{i,j}$ itself:

$$S = \sum_{l=i-1}^{i+1} \sum_{m=j-1}^{j+1} s_{l,m}^{n}. \qquad (11.6)$$

Each node evolves from one temporal iteration to the next, according to the following three (relatively) simple rules:

Rule 1. If a node is in the rest state ($s^n = 0$), its state at the next iteration is given by

$$s^{n+1} = \min\left(\frac{N_r}{r} + \frac{N_a}{a}, A\right). \qquad (11.7)$$

Rule 2. If a node is in the recovery stage ($0 < s^n < A$), its state at the next iteration is given by

$$s^{n+1} = \min\left(\frac{S}{N_r + 1} + g, A\right). \qquad (11.8)$$

Rule 3. If a node is activated ($s^n = A$), it transits to the rest state at the next iteration:

$$s^{n+1} = 0. \qquad (11.9)$$

Here r, a, and g are all positive constants, and the resulting numerical values for s^{n+1} are truncated to the nearest lower integer when computing Rules 1 and 2, since the state variable s is an integer quantity.

How do these rules relate to the excitation–recovery cycle of figure 11.1? First, the state variable s is to be associated with the recovery variable Y. Because of its truncation to the lowest integer, Rule 1 captures the activation threshold dynamics represented by the dotted arrow, with the numerical values of the parameters r and a setting the value of this threshold. This is a "fast" process, as it operates in a single temporal iteration, and the resulting value of s represents the postactivation state. The acceleration parameter g in Rule 2 sets the rate at which s grows once activated, i.e., it sets the upward climbing speed along the right edge of the phase space path. As long as $g \ll A$, this can be considered a "slow" process, in that many temporal iterations are required to travel up from the postactivation state to saturation. The activation threshold A is equivalent to the saturation value of Y. Rule 3 amounts to saying that the transition from this upper portion of the path back down to the rest state is "fast," i.e., it takes place in a single temporal iteration.

Consider first the behavior of a single node in the recovery phase $(0 < s^0 < A)$, surrounded by 8 inactive nodes $(s = 0)$; such a lattice state could only result from the initial condition, as Rule 1 above would normally preclude an isolated resting node from entering the recovery phase. But assuming such an initial state (s^0) can be prepared, with $N_r = 0$ and $S = s^0$ Rule 2 yields $s^1 = S/(N_r + 1) + g = s^0 + g$. Pursuing the iterative process we then have $s^2 = s^1 + g = s^0 + 2g$, $s^3 = s^0 + 3g$, etc. This describes linear growth of s^n, at a rate set by the value of g, which will continue until the activation threshold A is reached. The same behavior would characterize a group of neighboring nodes all sharing the same value of s, because then $N_r = 8$ and $S = 9s^n$, so that once again $s^{n+1} = s^n + g$; all nodes would grow linearly in time with slope g, activate in sync, and start growing anew from a value $s^n = 8/a$, as per Rule 1. The resulting cycle of recovery, activation, and return to the rest state results in a periodic sawtooth pattern similar to the nodal evolution in the OFC earthquake model (see figure 8.5) in the absence of redistribution by neighboring avalanching nodes.

One crucial difference with the OFC model, however, lies with the fact that in the hodgepodge machine, redistribution between nodes takes place not just when nodes are activating, but operates throughout the whole recovery phase, via the diffusive behavior built into Rule 2. With $g = 0$ and for a recovering node surrounded by other recovering nodes $(N_r = 8)$, Rule 2 becomes $s_{i,j}^{n+1} = S/9$, i.e., $s_{i,j}$ adopts the mean value of its neighborhood.[3]

---

[3]This is akin to linear (Fickian) diffusive processes, which in the steady state must satisfy Laplace's equation $\nabla^2 s = 0$; using centered second-order finite differences on a regular equidistant Cartesian grid, one can show that such

The diffusive behavior of the hodgepodge is illustrated in figure 11.2, displaying a succession of horizontal cuts through the middle of a 100×100 lattice, starting from an initial condition composed of a 20×20 block of nodes with $s = 250$ at the lattice center, and $s = 0$ everywhere else. This solution uses parameter values $a = r = 0.1$, $A = 255$, and $g = 0$. On this 1-D cut, the initial condition (in black) shows up as a rectangular shape which spreads laterally and flattens with time, adopting a Gaussian-like shape. This is exactly the behavior expected from classical linear (Fickian) diffusion. At the outer edge of this spreading structure, nodes initially with $s = 0$ are pushed into the recovery phase, producing a recovery front propagating outward at a speed of one node per iteration. Each resting node hit by this front finds itself with three neighbors in the recovery stage, and so jumps to a nodal value $N_r/r = 30$, as per Rule 1. Here, because $g = 0$ and all surrounding nodes have the same value $s = 30$, once pushed into the recovery state, nodes experience no further growth.[4]

Now there enters a nonzero acceleration parameter g. As soon as the recovery front hits a node, growth at a rate set by g begins. Once activated, each node is surrounded by other nodes either just activated or beginning their recovery phase, so all grow at essentially the same rate. However, because the front propagates outward at one node per iteration, each node lags its predecessor by one g-sized step in the growth process. The

---

steady-state solutions must satisfy

$$s_{i,j} = \tfrac{1}{4}(s_{i-1,j} + s_{i+1,j} + s_{i,j-1} + s_{i,j+1});$$

i.e., $s_{i,j}$ is equal to the average of its 4 nearest neighbors, top/down/left/right.

[4]Readers familiar with the modeling of diffusive processes may note some unexpected features in figure 11.2, particular in the late evolutionary phases. The lateral broadening of the central bumps seems to come to a standstill around iteration 100, after which slow inward shrinking ensues; this is not behavior expected of linear diffusion. The culprit is the truncation to the lowest integer applied to the computation of Rule 2, which effectively acts as a sink term, slowly "removing" chemicals from the system. In other words, diffusion in the hodgepodge machine is nonconservative.

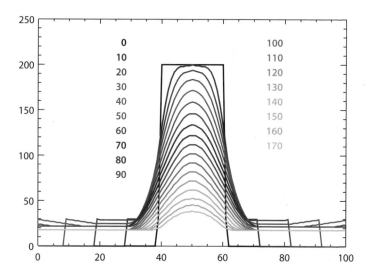

Figure 11.2. Diffusive behavior in the hodgepodge machine, for parameter values $r = 0.1$, $a = 0.1$, and $g = 0$. The initial condition is $s = 250$ in a 20×20 block of nodes at the center of a 100×100 lattice, and $s = 0$ elsewhere. The various color-coded lines are horizontal cuts through the middle of the lattice, plotted and color coded at a 10-iteration cadence, as shown.

presence of this systematic lag results in an outward-propagating sawtooth waveform, dropping to zero, and beginning anew when nodes reach the activation threshold, a direct reflection of the temporal sawtooth pattern locally characterizing the evolution of each node.

   Figure 11.3 shows four snapshots of a simulation with parameter values $a = r = 0.1$ and $g = 10$, now on a 200×200 lattice and starting from the same "central block" initial condition of figure 11.2. The top row of images shows a grayscale coding of the state variable s at iterations 50, 60, 70, and 80, going from left to right. The four spreading planar wave fronts emanating from the lattice center are quite obvious, and show curvature only near the intersections of their four phase fronts. The bottom plot shows

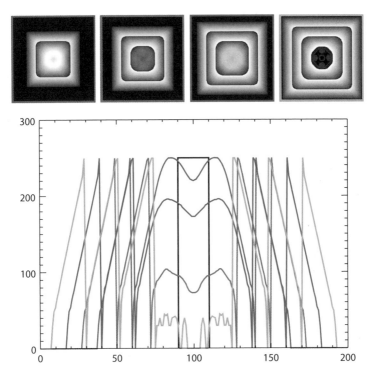

Figure 11.3. Wave generation and spreading in the hodgepodge machine, for parameter values $a = 0.1$, $r = 0.1$, and $g = 10$. The initial condition is $s = 250$ within a 20×20 block of nodes at the center of a 200×200 lattice, and $s = 0$ elsewhere. The four panels on top show a grayscale representation of $s_{j,k}$ at iterations 50 (framed in blue), 60 (purple), 70 (red), and 80 (green). The bottom panel shows the corresponding horizontal cuts along the center of the lattice, at the same four epochs, as color coded.

horizontal cuts across the lattice center, the outward-propagating sawtooth pattern being now most obvious.

For a propagating plane wave front, activation of resting nodes ahead of the front drives their state up to a value $s = 3/a$, after which they grow by an increment g at each iteration until they reach the activation threshold; if diffusion is neglected, the period of this activation–recovery cycle is then $(A - 3/a)/g \simeq 22$ for the parameter values used in figure 11.3. Because the front advances by one node per iteration, the corresponding wavelength of the propagating sawtooth wave is then 22 nodes, in agreement with the wave pattern observed in figure 11.3.

11.3 Numerical Implementation

Figure 11.4 offers a simple implementation of the hodgepodge machine in the Python programming language. The overall code structure closely resembles the OFC CA encountered in chapter 8 (see figure 8.3), the primary differences being at the level of lattice state updates (lines 34–45), which are carried out here according to equations (11.7)–(11.9), rather than the simpler equations (8.10)–(8.13). Both models use a random initial condition (lines 26–28) and enforce synchronous update of the lattice. Note however that periodic boundary conditions are used here, in contrast to the "frozen" boundary conditions of the OFC model. This is implemented via the user-defined function `periodic`, which operates on a 2-D array given as an argument (here named internally `grid`; see figure D.3) but does not return an explicit result. Note the use of Python's `elif` keyword, a contraction of the usual **else if** construct. Finally, because the state variable is an integer, the computations of Rules 1 (line 45) and 2 (line 41) truncate to the lowest integer via the Python integer conversion function `int`.[5] Likewise, the use of Python's `min` function ensures that $s_{i,j}^{n+1} \leq A$ even if a or r are set to very small values.

---

[5] I coded up the hodgepodge machine in C, IDL, and Python, and kept finding small but puzzling differences in some parts of the parameter space; they turned out to be related to the manner in which these various computing languages deal with truncation and conversion to integers. So be warned.

```
1   # PATTERN FORMATION BY THE HODGEPODGE MACHINE ON A 2D LATTICE
2   import numpy as np
3   import matplotlib.pyplot as plt
4   #---------------------------------------------------------------------
5   N =128                                  # Lattice size
6   AA=255                                  # Activation threshold
7   a =1.0                                  # Activation parameter
8   r =5.                                   # Recovery parameter
9   g =30.                                  # Acceleration parameter
10  n_iter=423                              # Number of iterations
11  #---------------------------------------------------------------------
12  # FUNCTION PERIODIC: enforces periodicity (see Fig D.3)
13  def periodic(N,grid):
14      grid[1:N+1,0]   =grid[1:N+1,N]       # Horizontal periodicity
15      grid[1:N+1,N+1]=grid[1:N+1,1]
16      grid[0,1:N+1]   =grid[N,1:N+1]       # Vertical periodicity
17      grid[N+1,1:N+1]=grid[1,1:N+1]
18      grid[0,0],grid[N+1,N+1]=grid[N,N],grid[1,1] # The four corners
19      grid[0,N+1],grid[N+1,0]=grid[N,1],grid[1,N]
20  # END OF FUNCTION PERIODIC
21  #---------------------------------------------------------------------
22  # MAIN PROGRAM
23  dx=np.array([-1, 0, 1,1,1,0,-1,-1])     # Template arrays
24  dy=np.array([-1,-1,-1,0,1,1, 1, 0])
25  state  =np.zeros([N+2,N+2],dtype='int') # Lattice array
26  for i in range(1,N+1):
27      for j in range(1,N+1):
28          state[i,j]=np.random.random_integers(0,AA)
29  periodic(N,state)                       # Enforce periodicity
30  for iterate in range(0,n_iter):
31      update=np.zeros([N+2,N+2],dtype='int') # Lattice update array
32      for i in range(1,N+1):                 # Main lattice loop
33          for j in range(1,N+1):
34              suma,sumr,nsum=0.,0.,1.*state[i,j] # Initialize counters
35              for k in range(0,8):             # Loop over nearest-neighbors
36                  ns=state[i+dx[k],j+dy[k]]
37                  nsum+=ns                     # Sum all nodes (Eq 11.1)
38                  if (ns == AA): suma+=1       # Sum activated nodes
39                  if (ns > 0) and (ns < AA): sumr+=1 # Sum recovering nodes
40              if state[i,j] > 0 and state[i,j] < AA: # In recovery phase
41                  update[i,j]=min([AA,int(nsum/(sumr+1.)+g)]) # Eq (11.3)
42              elif state[i,j] == AA:           # Enter rest phase
43                  update[i,j]= 0
44              elif state[i,j] == 0:            # Enter recovery phase
45                  update[i,j]=min(AA,int(suma/a+sumr/r))  # Eq (11.2)
46  # End of main lattice loop
47      periodic(N,update)                   # Enforce periodicity
48      state=update                         # Synchronous update
49  # End of temporal iteration
50  plt.imshow(state,cmap="gray",interpolation="nearest")
51  plt.show()
52  # END
```

Figure 11.4. A minimal implementation of the hodgepodge CA in Python.

11.4 Waves, Spirals, Spaghettis, and Cells

The operation of the hodgepodge machine combines a local activation–recovery cycle with spatial spreading and entry into the recovery phase mediated either by neighbor proximity or diffusion. These processes, as embodied in the hodgepodge machine, are not particularly complicated, yet they can lead to a staggering array of patterns and behaviors as the model's defining parameters are varied. Figure 11.5 shows four examples, in all cases starting from a random initial condition where the state nodal variable is drawn randomly at each node from the interval $[0, A]$. These four solutions are all computed on a 128×128 lattice, with $A = 255$ and other model parameters as listed over each snapshot, the latter all taken after 500 iterations.[6] Horizontal and vertical periodicity is enforced at the lattice boundaries.

The solution displayed in figure 11.5A (top left) produces irregularly shaped activation fronts propagating across an otherwise diffuse profile for the state variable. In this parameter regime, the hodgepodge machine behaves a bit like the forest-fire model of chapter 6 in some portions of its parameter space. One important difference here is that the evolution of any given node is quasiperiodic, with a mean periodicity of 42.6 iterations for this specific solution, a feature to which we shall return shortly. The low value of g implies that diffusion (Rule 2) dominates the evolution except in the immediate vicinity of an activation front.

In the solution displayed in figure 11.5B (top right) activation fronts are still present, but now propagate with a well-defined wavelength, much as in figure 11.3, and are organized spatially in

---

[6]The choice of color table can have a large impact on the structures visible when displaying the state variable as an 8-bit pixelated image, as in figure 11.5. The grayscale adopted here (direct grayscale for panels A and B, reverse grayscale for C and D) is the most neutral, but you can have fun with this by exploring the various predefined color tables that can be supplied as an optional argument to the matplotlib function `imshow` in the code of figure 11.4. Don't be afraid to follow your innate artistic impulses; have fun with it!

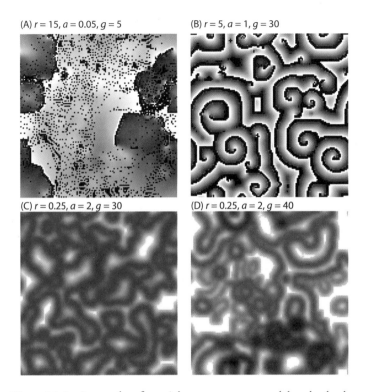

(A) $r = 15, a = 0.05, g = 5$

(B) $r = 5, a = 1, g = 30$

(C) $r = 0.25, a = 2, g = 30$

(D) $r = 0.25, a = 2, g = 40$

Figure 11.5. A sample of spatial patterns generated by the hodge-podge machine, starting from a random initial condition. All these simulations are carried out on a 128×128 lattice, with $A = 255$, and other parameter values as listed. The temporal recurrence periods for these solutions are $P = 42.6$, 9.6, 12.0, and 10.0 iterations respectively, going from panel A through D.

the form of spreading spiral waves with foci distributed randomly across the lattice. Geometrically intricate patterns are produced when spirals spreading from neighboring foci meet, with the wave fronts merging, interfering with, and annihilating each other. Some spiral foci occasionally disappear while others appear through fragmentation of existing spiral fronts interacting with one another. Production of new spirals often takes place from the tips of broken wave fronts, and both senses of rotation are equiprobable. The nodal recurrence period corresponds to the revolution period for the spirals, equal to 9.6 iterations for this solution.

Figure 11.5C (bottom left), displays an entirely different pattern, which is perhaps best described as thick overcooked spaghetti. No wavelike propagation is taking place here; instead the spatial pattern remains frozen as the nodal variable increases to the activation threshold; however, after the nodes activate, a new spaghetti pattern is produced, and yet another after the next activation cycle, which for this solution has a period of 12 iterations. There is a qualitative behavioral similarity here with the spatial domains developing in the OFC earthquake model (see figure 8.6), where the spatial shape of domains evolves only at their boundaries, from one avalanching cycle to the next.

The solution displayed in figure 11.5D (bottom right) evolves similarly, going through sequences of spatially steady patterns growing to activation, then reemerging with a new spatial distribution. For these parameter values, the pattern includes many large cell-like structures, some double walled, some with more intricate internal structure. In this part of the model's parameter space, solutions are sometimes encountered where only small cells are produced at first, and as the solution goes through successive collective activation cycles, one "supercell" with complex internal structure slowly takes over the domain, only to later disintegrate again into small cells, this long spatiotemporal quasi-cycle then beginning anew.

The four solutions displayed in figure 11.5 sample only a small subset of spatial patterns that can be produced by the hodgepodge machine. Other types of spatial patterns include diffuse cloud-like structures, structured binary noise, and mixtures of homogeneous and inhomogeneous regions; and that is without even playing with the threshold parameter A or nearest-neighbor template! Moreover, in many parts of parameter space, the hodgepodge

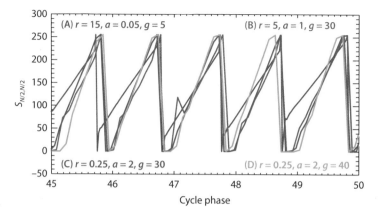

Figure 11.6. Time series of the state variable s sampled at the center of the lattice, for the four simulations of figure 11.5. The time series are plotted in units of cycle phase, and a phase offset has been artificially introduced so that all peaks line up. These time series closely resemble each other, even though the spatial patterns they produce do not (see text).

machine also shows sensitivity to the choice of initial condition. All this complexity arises in a CA defined by four primary numerical parameters. How can this be?

As a first step toward answering this question, now consider figure 11.6, showing time series segments of the state variable s^n for a node located at the lattice center, for the four hodgepodge simulations of figure 11.5. These four solutions have recurrence periods increasing with decreasing values of g, as per our earlier discussion, so that in constructing this plot, "time" (measured in iterations) is divided by the recurrence period of each solution. The horizontal axis becomes a measure of cycle phase, and on such a plot, all solutions have a mean period of unity. All four time series exhibit basically the same sawtooth pattern, namely, gradual, quasi-linear growth to the activation threshold, followed by a rapid, single-iteration drop to the rest state once this threshold is reached. Solutions with low values of r show some curvature or even spikes at the beginning of the recovery phase, as a consequence of the rapid upward jump produced by equation (11.7), Moreover, the solutions are not strictly periodic, as is readily noted upon examination of figure 11.6 for solution A (in blue), and D (in green), the latter also exhibiting significant variations in the duration of the resting phases.

Still, how can the nodal time series be so similar, when the spatial patterns of the solutions displayed in figure 11.5 are so different? We should first note that because the recurrence cycle periods are not the same for the four solutions of figures 11.5 and 11.6, diffusion has more time to operate during the recovery stage of the longer cycle solutions than in their more rapidly cycling cousins. The most important factor, however, is the relative *spatial phase* of neighboring nodes: by how much is each node lagging each of its eight neighbors in the activation–recovery cycle, and does this lag have any directional bias? Much insight into these questions can be obtained by comparing and contrasting planar and spiraling wave fronts, the exercise to which we now turn.

11.5 Spiraling Out

Spirals are arguably the most visually striking and intriguing patterns produced by the hodgepodge machine. They have also attracted the most attention, because spiral waves are observed in many types of excitable systems. These include the Belousov–Zhabotinsky reaction and other similar chemical reaction–diffusion systems, but also biological systems such as slime molds and starving amoeba colonies. It has also been suggested that some classes of cardiac arrythmia could be associated with the breakup of the electrical wave fronts normally propagating across the heart muscles into localized spiral waves, induced by tissue damage. The remainder of this chapter thus focuses on understanding the generation of spiral waves in the hodgepodge machine.

It will prove useful first to go back to the planar wave fronts of figure 11.3; more specifically, let's focus on the vertically oriented planar wave front propagating to the right in the sequence of four snapshots. Except near the corners of the expanding square wave front, nodes connected in the direction parallel to the wave front all cycle in phase. Phase differences materialize only between nodes in the propagation direction of the planar wave front. This occurs because each node has two neighbors (above and below) sharing the same value of the state variable, three having the same higher value (on the left), and another three the same lower value (on the right); this lateral ordering reverses only at activation. The hodgepodge rules then ensure that vertical invariance is preserved, and the same, of course, holds for horizontal invariance in the vicinity of vertically propagating plane wave fronts. Note that the square form of the spreading wave is not set by the square pattern of the initial condition used to generate the solution displayed in figure 11.3. For these parameter values, activity propagates one node per iteration also along diagonals; in other words, in terms of geometrical distance, a planar activation front inclined by 45 degrees with respect to the lattice grid lines propagates faster than horizontal or vertical wave fronts, by a factor $\sqrt{2}$. This implies that the circular wave front initially produced by a circle-shaped initial condition will inexorably evolve into a square spreading wave.[7] Diffusion, on the other hand, tends to smooth out gradients, and so it will tend to turn sharp corners into curved arcs. The persistence of curved wave fronts thus reflects a balance between propagation (Rule 1) and diffusion (the diffusive part of Rule 2).

Consider now a node in the rest state ($s = 0$), located just behind a propagating planar activation wave front ($s = A$); such

---

[7]A similar squaring of burning fronts takes place in the forest-fire model of chapter 6, when the density of trees through which the burning front moves is sufficiently high.

a node just entered the rest state at the preceding iteration. If the parameter $a \leq 3$, then Rule 1 will push it into the recovery state at the next iteration (remember that Rule 1 truncates to the lowest integer). If, on the other hand, $a > 3$, then the node will stay in the rest state, and the lattice will remain forever inactive after the passage of the wave front, unless diffusion from elsewhere is efficient enough to trigger entry into the recovery phase. This latter situation is akin to the radial spread of the epidemic wave front in figure 9.6, behind which no surviving agents remain, so that the epidemic cannot "reactivate" behind the front unless enough healthy random-walking agents stumble their way back into the infected area.

All of this becomes more interesting if, for whatever reason, the wave front breaks. Nodes located behind the last active node of the wave front, and having just entered the rest state, may now have neighbors that are in the recovery stage ($0 < s < A$), in which case Rule 1 can lead to reactivation provided r is small enough. The effect will be to extend the wave front beyond its original tip, but this extension will lag in time (unless $g \sim A$), meaning that it will curl back inward toward the region located behind the bulk of the planar wave front, eventually leading to reactivation in those regions. This is the mechanism leading to the development of spiral waves in some regions of the hodgepodge machine's parameter space.

Figure 11.7 shows a close-up of the core of one of the spirals developing in the simulation displayed in figure 11.5B. These 10 frames span one revolution of the spiral. However, the bottom-right snapshot is not quite identical to the top left. This is because the recurrence period (see figure 11.6) for this solution is 9.6 iterations rather than 9.0. Examine closely the evolution of the activation front (in red) in the core of the spiral, and see in action the process of wave-front extension and curling just described. In particular, notice how the inside end of the radially expanding activation front always grows toward its

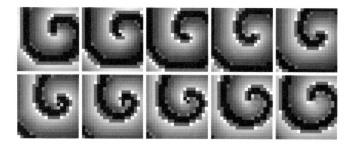

Figure 11.7. Close-up of a spiral core spanning a full revolution of the spiral. Frames are spaced one temporal iteration apart, with time running from left to right and top to bottom. Active nodes $(s_{i,j}^{n} = A)$ are colored in red, with black corresponding to resting nodes $(s_{i,j}^{n} = 0)$ and the grayscale spanning the recovery range $[1, A - 1]$, from dark to light. Parameter values are as in figure 11.5B.

left (as measured with respect to its local, approximately radial, propagating direction), into a region containing recovering nodes approaching the saturation threshold (light gray).

By the above logic, a planar wave-front segment should curl inward at both ends, and one can imagine the curling ends eventually meeting and regenerating a new planar wave front. Such a system can be viewed as a pair of counterrotating spiral cores. Figure 11.8 shows the evolution of two such pairs interacting with one another. Near the center of the first frame (top left), a short, approximately planar, activation front is propagating toward the bottom-right corner. The curling back inward of its tips is clearly apparent in the subsequent five frames, persisting until the upper tip merges with the activation front generated by another spiral core. The merging produces a new, approximately planar wave front, propagating downward (frames 7 to 10) until another merging event with the lower curling tip of the first wave front finally regenerates the original wave front propagating toward the bottom-right corner.

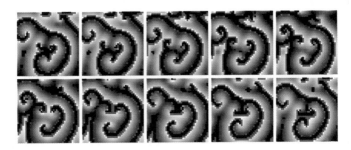

Figure 11.8. Interaction of spiral waves in the same simulation as in figure 11.7. The interacting spiral waves are generated here by two pairs of counterrotating cores. Parameter values are as in figure 11.5B. Compare to the shape of burning fronts in figure 6.5.

This curling back of activation wave fronts can actually be observed in other systems investigated in some of the preceding chapters: they materialize in some parts of parameter space for the forest-fire model of chapter 6 (see figure 6.5), as well as in the epidemic spread model of chapter 9. As the saying goes, finding these regions of parameter space is left as an exercise!

Take a last look at figure 11.5B; there are pretty much as many spirals rotating clockwise as counterclockwise. The locations of the cores and the sense of rotation of these spirals are determined by the specific realization of the random initial conditions. In a given region of the lattice, the spatially extended patterns reflect the action of the hodgepodge machine's dynamical rules working off this initial condition. These rules are isotropic, implying that nothing in their formulation favors one sense of rotation over another; the emergence of two senses of rotation is another instance of spontaneous symmetry breaking.

11.6 Spontaneous Pattern Formation

The formation of statistically stable, persistent patterns from a random initial condition represents yet another instance of order emerging from disorder. Equilibrium thermodynamics does not allow this, so the explanation of pattern formation by the hodgepodge machine must again be sought in terms of open dissipative systems.

How can the hodgepodge machine be deemed "open" and "dissipative"? The dissipative aspect is related to the nonconservative nature of the diffusive process embodied in equation (11.8), as already discussed in relation to figure 11.2. The "open" aspect is harder to pinpoint, and its exact form depends on the nature of the excitable system under consideration. For the hodgepodge machine, it is hidden in the manner in which the activation–recovery cycle of figure 11.1, involving the two dynamical variables X and Y, has been reduced to tracking a single quantity (s) related to the recovery variable Y, whose evolution is determined by rules defined with *fixed* numerical values for parameters r, a, and g. This implies an external regulatory mechanism that maintains constant operating conditions for the system, i.e., the system is not closed.

Notwithstanding such interpretative subtleties, it remains quite remarkable that the wonderful array of spatiotemporal patterns produced by the hodgepodge machine results simply from coherent spatial variations in the phase of the nodal recurrence cycle of neighboring nodes. Any one single node does the same thing as its neighbors: activates, grows slowly to saturation, and then falls back to the rest state, and with a cycle period that is the same for all nodes. The spatial phasing leading to pattern is established and sustained by the interplay of threshold-based excitation, growth, and diffusive local spreading. The latter being in essence the macroscopic manifestation of a microscopic random walk (see appendix C.6), the hodgepodge machine is truly producing (large-scale) order out of (small-scale) disorder, not just via the initial condition, but also via its underlying "microscopic physics."

11.7 Exercises and Further Computational Explorations

1. Similarities between behaviors observed in the hodgepodge machine and the forest-fire model of chapter 6 have been noted repeatedly in this chapter. Try to find values of the hodgepodge parameters a, r, g, and A that best mimic the behavior of the forest-fire model in the limit where p_g is (relatively) high and p_f is very small (see, e.g., figure 6.5).

2. Repeat the simulations of figure 11.5 using the following initial conditions:

   a. a circular block of nodes with $s^0 = 250$ sitting at the lattice center;

   b. a thick line segment (a 10×100) block of nodes $s^0 = 250$ sitting at the lattice center;

   c. a few one-node-wide straight lines of $s^0 = 250$ nodes set at random angles with respect to the lattice grid lines (these lines are allowed to intersect).

   How dependent is the behavior of the hodgepodge machine on the initial conditions?

3. Our discussion of wave propagation in the hodgepodge machine simulations of figure 11.3 has not considered the effect of diffusion, and the aim of this exercise is to do just that.

   a. Repeat the simulation of figure 11.3 for smaller and larger values of g. Is wave propagation always possible? How are the wavelength and wave propagation speed varying with g (keeping $a = r = 0.1$)?

   b. Diffusion can be eliminated altogether from the hodgepodge machine by replacing Rule 2 by $s^{n+1} = \min(s^n + g, A)$. Repeat your previous set of experiments with this diffusionless Rule 2. How are your results altered?

4. Using the same hodgepodge parameter values as in figure 11.5B, design an initial condition that produces a single spiral with its core at the lattice center. How can you control the spiral's direction of rotation?

5. Explore the behavior of the following two variants of the hodgepodge machine:

   a. Redefine Rules 1 and 2 so that only the closest four neighbors (top+down+right+left) are involved.
   b. Redefine Rules 1 and 2 so that they involve a more spatially extended neighborhood, namely, all nodes in the range $(i \pm 2, j \pm 2)$ of node (i, j), namely, 24 neighbors, with the same weight given to each.

   Can you still produce spiral waves under these setups? What about simple and/or complex cells?

6. The Grand Challenge for this chapter is a real bear, in fact bordering seriously on a true research project: generalizing the hodgepodge machine to three spatial dimensions. The required coding developments are straightforward, and, fundamentally, the behavior of the 3-D hodgepodge machine is still defined by the same four parameters a, r, g, and A as in its 2-D cousin. However, visualizing results pretty much requires some skills (or learning effort) in 3-D data rendering and visualization. Explore the spatial patterns produced by the 3-D hodgepodge machine for varying parameter values. If you manage to produce double-coiled helixes, let someone know because you may be on to something big!

11.8 Further Reading

An engaging and accessible discussion of excitable systems can be found in chapter 3 of

Goodwin, B., *How the Leopard Changed Its Spots*, Simon & Schuster (1994).

Autocatalytic chemical reactions and reaction–diffusion equations are discussed in numerous mathematical biology and chemistry textbooks, for example,

Murray, J.D., *Mathematical Biology*, Berlin: Springer (1989).

Specifically on the Belousov–Zhabotinsky reaction, I found the following article very informative:

Zhabotinsky, A.M., "Belousov–Zhabotinsky reaction," *Scholarpedia*, **2**(9), 1435 (2007).

This is available online, open access (March 2015):

http://www.scholarpedia.org/article/Belousov-Zhabotinsky_ reaction.

Turing's groundbreaking 1952 paper on pattern formation in reaction–diffusion systems still makes for a fascinating read; it is reprinted in chapter 15 of

Copeland, B.J., ed., *The Essential Turing*, Oxford University Press (2004).

On the hodgepodge machine, see

Gerhardt, M., and Schuster, H., "A cellular automaton model of excitable media including curvature and dispersion," *Science*, **247**, 1563–1566 (1990);

Dewdney, A.K., "The hodgepodge machine makes waves," *Scientific American*, **225**(8), 104–107 (1988).

12

EPILOGUE: NATURAL COMPLEXITY

There *are* things to hold on to. None of it may look real,
but some of it is. Really.
— (THOMAS PYNCHON, *Gravity's Rainbow*)

What I cannot create, I do not understand.
— (R.P. FEYNMAN, 1988)

12.1 A Hike on Slickrock

This is far from our first hiking trip in southeastern Utah, but
this one Easter trip has a new twist to it: our thirteen-year-old
son, an avid unicyclist, has taken his mountain unicycle along
to test his skills on the world-renowned slickrock mountain bike
trails of the Moab region. Day two finds us parked up Sand
Flats Road, on the barren plateau overlooking the Colorado River
and Moab Valley, at the trailhead of the legendary slickrock
loop.

This place is a burning hell in summer months, but in early
April it makes for quite a pleasant hike, with impressive views
down into the surrounding canyons. But up where we are it is
really pretty much all slickrock, and the few small trees and shrubs
are few and far between. Grasses, cactuses, and wildflowers do
manage to grow here and there, in patches of soil and debris
having accumulated in cracks and shallow depressions in the

rock, but by far the most common biological presence, besides mountain bikers, is lichen.

Lichens are one of the earliest and most successful symbiotic experiments of the biological world. Lichen is really algae and a fungus teaming up in a mutually beneficial relationship; the algae make food through photosynthesis, while the fungus provides structural support and anchoring, and gathers moisture and nutrients from the environment. The deal works, and very well, as varieties of lichens are found in the most extreme environments, from the arctic tundra to the driest deserts.

The desert environment is indeed very harsh, and most lichens I see on the rocks look pretty dried up, and, I'm guessing, are long dead. I don't know much about lichens, but I'm presuming growth takes place mostly in the spring, while the porous sandstone surface still holds some moisture and the sun is not yet scorching the rocks. I have since learned that the lichens I am seeing belong to the family of crustose lichens, which usually grow radially outward on their substrate. I do see plenty of more or less circular patches of varying colors and sizes. I also see lichen rings. Upon examination, it just looks like the central part has dried up, died, and flaked off, leaving a ring-shaped structure. It does makes sense. In some cases regrowth has taken place inside an existing ring, presumably in a later wet season, leading to a pattern of concentric irregular rings. This makes sense also, I'm guessing. Figure 12.1 shows some particularly nice examples, captured in the Fiery Furnace area of Arches National Park.

But what really catches my attention are the spirals. They may not be the most common pattern characterizing the growth of crustose lichen on slickrock, but they show up often enough, in different types of lichens, on different types of rocky surfaces inclined at widely varying angles with respect to gravity or the noon sun. The more spirals I see, the more I see a pattern in there, something *robust*. As we make our way across the rocky landscape, I find myself pointing my camera to the ground

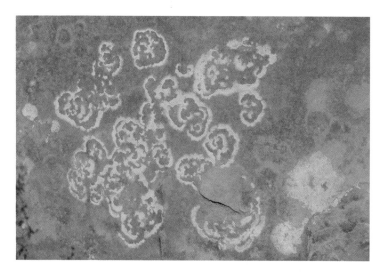

Figure 12.1. Crustose lichen on the desert slickrock of southeastern Utah. The true vertical dimension of the image is about 40 centimeters. Photograph by yours truly.

with increasing frequency. Fortunately for me, passing mountain bikers are too awestruck at our son careening up and down the double-black-diamond mountain bike trails on his unicycle to become concerned enough with my combination of foreign accent and odd photographic behavior to dial the Homeland Security hotline.

The top image in figure 12.2 shows an example of some of the spiral-shaped patterns I photographed. I soon start to notice instances of double-spiral-like structure, such as in the middle and bottom photographs in figure 12.2, where the growth front curls back inward symmetrically about some bisecting axis, morphologically similar to those generated by the hodgepodge machine in the "spiral" region of its parameter space (see figure 11.8). A few such structures are also visible in figure 12.1, if you look carefully.

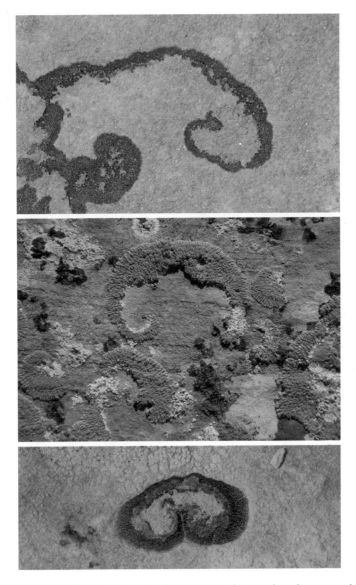

Figure 12.2. Examples of spiral patterns and inward-curling growth fronts in various types of crustose lichens in southeastern Utah. Compare to figure 11.8. Photographs by yours truly.

How can this be? Lichen growth requires moisture, but also depletes water from the rock's surface. There is probably some nonlinearity in there somewhere. Can water be considered the activation variable, and lichen growth a recovery variable tracing an activation–recovery cycle? Or would it be the other way around? Looking closely at the rock texture and color ahead of and behind the lichen "front" reveals a definite asymmetry, faintly visible in the top photograph in figure 12.2, especially around the spiral core. This looks very much like what the recovery variable does on either side of the wave fronts produced by the hodgepodge machine.

By then my mind is racing, dreaming up excitable systems and reaction–diffusion dynamics. I am well aware that I am engaging in a dangerous exercise, namely, forcing a known explanation on an intuitive hunch; but the visual evidence seems just too compelling for this to be a morphological convergence without any common dynamical origin. I sure wish I knew more about lichen growth.

That same evening, watching the sunset over Canyonlands from the porch of our rental cabin halfway up the La Sal mountains, it slowly dawns on me that my spontaneous and freewheeling speculations on lichen growth and form have strangely paralleled an experience lived centuries ago by another physicist, also looking for a break, and also out for walk.

12.2 Johannes Kepler and the Unity of Nature

History has not recorded the exact why or when, but one morning in the winter of 1609, Johannes Kepler decided to take the day off. For more than a decade he had labored relentlessly to produce a working model of planetary orbits from the store of unprecedentedly accurate astronomical observations of Tycho Brahe. He had arrived in Prague 10 years before, to be employed as Brahe's senior assistant. Following the untimely death of his boss in 1601, Kepler finally secured full and unrestricted access to the needed data, as well as inheriting the job of Imperial Mathematician to Emperor Rudolph II. Professionally secure and, perhaps more importantly, freed from pressures to vindicate Brahe's pet planetary model, Kepler embarked on a computational effort that would overthrow basic astronomical tenets that had endured for over two millennia.

Today, Kepler is remembered primarily for having deduced from Brahe's observations the three laws of planetary motion that bear his name. This may appear entirely in line with astronomical tradition, which up to the times of Kepler and Galileo had primarily sought accurate mathematical descriptions of planetary motion. In reality, Kepler did break from astronomical tradition, perhaps even more so than Galileo, by seeking *physical causes* for the numbers, motion, and overall arrangement of the six solar-system planets known at the time. His writings, very much in the stream-of-consciousness style and often veering into downright geometrical mysticism, do not exactly make for easy reading today, and I suspect must also have baffled many an early seventeenth-century astronomer.

Already, in his 1596 book entitled *Mysterium Cosmographicum*, Kepler had put forth a daring hypothesis relating natural order to geometry. His idea was that the relative sizes of planetary orbits could be deduced from the nesting pattern of the five regular platonic solids. In later years he went on to consider the relation of planetary orbital periods to the frequency ratios of musical harmonies, and even the possibility that a magnetic field emanating from the rotating Sun was responsible for carrying the planets along their orbits. These ideas may appear naive in retrospect, but they do reveal a sharp and inquisitive mind bent on *explaining* astronomical facts, in the most modern sense of the word.

Who knows what Kepler was actually thinking about when he started walking the streets of Prague on that winter morning in 1609. But what was initially a casual walk soon took an unexpected turn. Kepler himself later described the event:[1]

> Just then by a happy chance water-vapour was condensed by the cold into snow, and specks of down fell here and there on my coat, all with six corners and feathered radii. Upon my word, here was something smaller than any drop, yet with a pattern. (p. 7)

Being the astronomer that he was, he marvels that "it comes down from heaven and looks like a star." Figure 12.3 shows photographs of snowflakes having formed under varying meteorological conditions. No single snowflake is ever exactly like another, and there exists, to all intents and purposes, an infinity of shapes intermediate between the thin six-pointed "needle-star" (top left) and the solid hexagonal plates (bottom right).[2]

Marveling at the delicate shapes of snowflakes on his sleeves, Kepler rapidly notices that all the single snowflakes he observes are planar structures harboring six highly similar branches. He immediately formulates an absolutely typical Keplerian question: *Why six?*, which is soon joined by another: *Why flat?* Kepler, an accomplished mathematician, goes on to consider the close packing of spherical water droplets in the plane, noting that the resulting hexagonal pattern has the same sixfold symmetry as his snowflakes. This could, in principle, "explain" both the observed planar structure and the symmetry. Despite it being firmly anchored in geometry, for Kepler this is not an appropriate *physical* explanation. He wants to know what *drives* this orderly assemblage of water droplets, and no other, upon condensation and freezing. Kepler argues that this organizing principle (he calls it *facultas formatrix*) cannot reside in the water vapor, which is

---

[1] All quotations are taken from the English translation of Kepler's 1611 booklet entitled *The Six-Cornered Snowflake*, listed in the bibliography to this chapter.

[2] This would be a good time to go back and take another look at figure 2.7!

Figure 12.3. Photographs of snowflakes having formed under varying atmospheric conditions. All these snowflakes are planar, except for the columnar crystal at the bottom left (seen here in side view). Photographs by Ken Libbrecht, used by permission.

diffuse and shapeless, nor can it be found in the individual water droplets themselves, which are spherical and unstructured. Kepler goes on to consider critically a number of working hypotheses, but rejects them one after the other as inadequate, to finally conclude with a daring statement, grounded in his profound belief in the unity of Nature:

> The cause of the six-sided snowflake is none other than that of the ordered shapes of plants and of numerical constants.... I do not believe that even in a snowflake this ordered pattern exists at random. (p. 33)

Today we understand that the sixfold symmetry of snowflakes is a reflection of the crystalline assemblage of water molecules in horizontally offset planar layers such that the oxygen atoms define the vertices of space-filling tetrahedra. The resulting assemblage of oxygen and hydrogen atoms in this crystal lattice happens to be the configuration that minimizes the free energy of the system. There you go. Under most meteorological conditions, growth occurs preferentially at the edges of the planar layers, rather than perpendicular to them, thus explaining the 2-D shape of (most) snowflakes.[3] However, the manner in which the 2-D growth takes places is influenced by surface diffusion along the outer planar surfaces of the growing crystal, and turns out to exhibit a very sensitive dependence to air temperature. Laboratory experiments have shown that changes of as little as 1°C can trigger, for example, a transition from solid hexagonal snowflakes to six-branch dendritic crystals. It is quite sobering to reflect upon the fact that, more than four centuries after Kepler's pioneering foray into crystallography, the morphogenesis of the common snow crystal is still not adequately understood in quantitative physical terms.

---

[3]In some temperature ranges, snowflakes grow as prismatic columns of hexagonal cross section, often capped at each end by a wider hexagonal plate; see the bottom-left panel in figure 12.3.

Since Kepler rejected atomism (in part for religious reasons), he would find the above explanation for the flatness and sixfold symmetry of snowflakes profoundly shocking, even though at the end of his 1611 book on the topic he presciently defers the explanation of snow crystals to "the attention of metallurgists and chemists." However, the atomistic grounding of the modern view of snowflake structure would likely not have been Kepler's strongest objection. At the end of his concluding essay accompanying the 1966 English translation of Kepler's book on snowflakes, Lancelot Law Whyte cogently encapsulates the most fundamental aspect of Kepler's views on the unity of Nature by formulating, in contemporary physical language, a question Kepler himself would certainly have approved of for hitting the nail right on the head:

> We should not expect complete knowledge of highly complex systems, but it is reasonable to require of science a simple explanation of simple observations. If the hexagonal snowflake is highly complex, is there no shortcut from the postulates of physics to our visual observations? What in the ultimate laws produces visually perfect patterns? (p. 63)

Whyte aptly entitled his concluding essay "Kepler's unsolved problem"; I could well have done the same with this chapter, because another fifty years later Kepler's problem is still not solved, but nowadays is considered to belong to the realm of the science of complexity.

The parallel between my hike on slickrock and Kepler's morning walk in Prague could be brought to a didactic climax if I were now to state that the said hike is what motivated the writing of this book; but this would be a lie. My interest in complex systems originates further back in time, with a physical phenomenon truly extraterrestrial: solar flares.

12.3 From Lichens to Solar Flares

Solar flares are the manifestation of extremely rapid and spatially localized release of magnetic energy in the extended atmosphere of the Sun, known as the corona. Because they can generate copious emissions of highly energetic radiation and relativistic beams of electrically charged particles which can pose a threat to astronauts and even space hardware, their prediction is a priority in the developing discipline known as *space weather*. Figure 12.4 shows an example of a large flare, viewed here in the extreme ultraviolet domain of the electromagnetic spectrum. This electromagnetic radiation is invisible to the eye and—fortunately for all of us surface-dwelling life forms—is completely absorbed in the very high atmospheric layers of the Earth. The image in figure 12.4 was captured from space, by the Extreme-ultraviolet Imaging Telescope (EIT) onboard the Earth-orbiting Solar and Heliospheric Observatory (SoHO). The flare causes the very bright EUV emission seen close to the solar limb on the right. Fainter emission is also seen all over the solar disk, often in the form of filamentary, loop-like structures extending above the solar surface. These trace out lines of force of the Sun's magnetic field, which structures the otherwise diffuse coronal plasma. Kepler was actually right about the Sun having a magnetic field extending into interplanetary space!

The pattern of ultraviolet emission in figure 12.4 is certainly complex in the visual sense, but there is more to it than that. The "size" of a flare can be quantified through the total energy released over the course of the event, which can be inferred from observations such as figure 12.4. Flare sizes span many orders of magnitude in energy release, and turn out to be distributed as a power law, with a logarithmic slope that is independent of overall solar activity levels, and is the same as inferred from flare-like emissions observed in stars other than the Sun. There is by all appearances something universal in flare energy release,

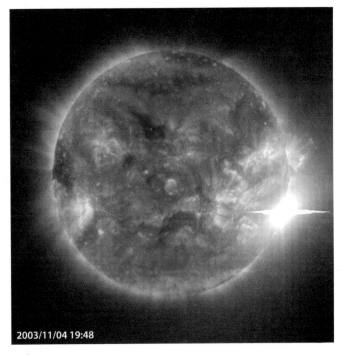

2003/11/04 19:48

Figure 12.4. A large solar flare (X28 in the NOAA classification scheme) observed by the EIT instrument onboard the solar observing satellite SoHO, a joint mission between NASA and the European Space Agency. The image shows radiative emission in the extreme ultraviolet (EUV), at a wavelength of 195Å. This is a false-color image, in which the intensity of EUV emission in each pixel is arbitrarily assigned a shade of green from a predefined color table (solar flares are not particularly green!). This flare, which occurred near the solar limb on November 4, 2003, is in all likelihood the strongest ever observed in the space era; we don't know for sure because the EUV emission was so intense that it saturated the imager, as evidenced here by the white horizontal streaks extending right and left of the flaring area.

something that is not sensitively dependent on details. Does this start to sound familiar?

It did to Edward Lu, a fellow postdoc in the early 1990s at the High Altitude Observatory of the National Center for

Atmospheric Research in Boulder, Colorado. Already well versed in flare physics through his doctoral research, Ed saw a connection with SOC and sandpile models, which at the time were spreading like wildfire in statistical physics. Teaming up with Russell Hamilton of the University of Illinois, the pair developed a three-dimensional sandpile model, in essence similar to that introduced in chapter 5. They identified the nodal variable with the coronal magnetic field, and used the curvature of the nodal variable, rather than the slope (or gradient) to define a stability criterion (see equation (5.3)). This choice was physically motivated, as it could be related to electrical currents induced by the stretching and bending of magnetic field lines, already known to be conducive to the triggering of a dynamical plasma instability known as magnetic reconnection. The latter was captured through simple but physically motivated local conservative redistribution rules, conceptually equivalent to equation (5.4), but differing in detail. Lu and Hamilton could show that upon being subjected to slow random forcing, much like in the simpler sandpile model of chapter 5, magnetic energy release occurs in the form of scale-invariant avalanches, characterized by a power-law size distribution with a logarithmic slope, comparing favorably to observations.

I remember very well Ed's enthusiasm at the time, and how hard he tried to "sell" his model to other flare researchers, not to mention funding agencies. Unfortunately he was too far ahead of his time, and the response he received all too often was along the lines of "it's interesting, but is it really physics?" The idea did percolate slowly through the field in the following decade, by which time many a solar physicist had followed in Ed's trailblazing footsteps, and many more have since. In the meantime, Ed had become an astronaut and was personally experiencing space weather on NASA's space shuttle and on the International Space Station. So it goes.

Such avalanches of magnetic reconnection events, if that is really what flares are, are not restricted to the Sun and other stars.

Large flares, such as that shown in figure 12.4, are often accompanied by the ejection of magnetized coronal plasma. These ejecta travel through the interplanetary environment, plowing up the solar wind along the way. Upon impinging on the Earth's magnetosphere, they trigger geomagnetic storms, the most spectacular manifestation of which being auroral emission, i.e., northern (and southern) lights. Substantial auroral emission also accompanies the so-called geomagnetic substorms, spontaneous and scale-invariant energy release events originating in the Earth's magnetotail, without any obvious solar trigger. It appears that substorms are closely akin to solar flares, in that they are driven by similar processes of magnetic field-line stretching and bending, leading to avalanches of spatially localized destabilization and magnetic energy release. Scale-invariant energy release is also observed in a number of more exotic astrophysical objects such as cataclysmic variable stars, pulsars, blazars, and accretion disks around black holes. Self-organized critical avalanche-type models for these objects have been developed, and offer an attractive explanatory framework for their pattern of energy release. In all cases, these are instances of natural complexity on the grandest of scales.

12.4 Emergence and Natural Order

Snowflakes, plants, arithmetically and geometrically significant numbers: Kepler had no qualms about assuming that inorganic, organic, and even mathematical systems share some common fundamental organizing principles. Running implicitly through this book is an assumption somewhat akin to Kepler's, in that similarly structured, simple computational models, all ultimately based on large numbers of elements (or agents) interacting through (usually) very simple rules, can capture emergent natural phenomena and processes as diverse as solar flares, avalanches, earthquakes, forest fires, epidemics, flocking, and so on. Rules at the microscopic level are simple; patterns and behaviors at the macroscopic level are not. How do we bridge the gap between the microscopic and the macroscopic? And under which conditions can the latter be reduced to the former?

Understanding—and even predicting—the behavior of a macroscopic system, on the basis of the physical rules governing the interactions of its microscopic constituents, has been achieved successfully in many cases. For example, one of the many great successes of nineteenth-century physics is the reduction of thermodynamics to statistical mechanics. Macroscopic properties of gaseous substances, such as pressure and temperature, as well as their variations in response to external forcing, can be calculated precisely if the nature of the forces acting between individual atoms or molecules of the gas are known. Even entropy, the somewhat esoterical thermodynamical measure of disorder in a macroscopic system, can be unambiguously related to the number of microstates available in the phase space of the system's microscopic constituents. Here, the microscopic rules are simple, and lead to simple "laws" at the macroscopic level—even though the intervening physico-mathematical machinery may not be so simple!

However, and even within physics, which deals typically with systems far simpler than organic chemistry or biology, this reductionist program often fails. Knowing everything about the quantum physics of a single water molecule H_2O would already be one tough Grand Challenge in an advanced graduate course on quantum mechanics; yet this microphysical knowledge, in and of itself, would be of little help in understanding why water flowing down a stream breaks into persistent swirls and vortices. What is it, lurking somewhere between the microscopic and the macroscopic, that evades reductionism?

Leaving the realm of physics, things rapidly get a lot worse, and we might as well jump immediately to what is arguably the most extreme example. Neurophysiologists are still a long way from understanding the working details of a single neuron, but even if they did, I don't think anyone would ever claim that a single neuron can "think." By all appearances, a great many neurons are required, and what seems to matter most is not so much the neurons themselves, but rather their pattern of synaptic interconnections. Still, can the 10^{14}–10^{15} interconnections of the 10^{10}–10^{11} neurons in the human brain explain consciousness? How many water molecules does it take to make a waterfall? Are these two questions really one and the same? Is it just, somehow, a matter of sheer numbers?

The "spontaneous" appearance of complex macroscopic behaviors irreducible to microscopic rules is now usually referred to as *emergence*. One can certainly argue that if the arising macroscopic behavior is unexpected, it simply means that we did not *really* understand the consequences of our imposed microscopic rules. In my opinion, writing off emergence in this way would be a spectacularly misguided instance of throwing away the baby with the bathwater. As simple as the computational models explored throughout this book may be, they do capture perhaps the essence of that elusive emergent something, that sometimes happens somewhere between the microscopic and macroscopic. Understanding that "something" is what the science of complexity is really about. When emergence has been explained, complexity will have been explained also.

Emergence is, almost by definition, a nonreductionist concept. Understanding it may require new ways to formulate questions and assess answers. Whether it is really "a new kind of science" is a matter of opinion. I have more than a few colleagues who would still today reply, "it's interesting, but is it really physics?" As far as I am concerned, it still fits comfortably within my preferred definition of science as *a way of knowing*.

12.5 Into the Abyss: Your Turn

So, what is complexity? I opened chapter 1 of this book by promising to keep clear of any formal definition of complexity, and I will resolutely stick to my word. My hope remains that by working your way through this book, coding up and running the various models for yourself, and trying your hand at the computational exercises and Grand Challenges, you have learned something useful and are coming out of it better equipped to tackle systems that are even more complex. There is certainly no lack of them all around us in the natural world.

The science of complexity is still young, and its future remains wide open. I do believe that it has something vital to contribute to humankind's most fundamental interrogations on the origin of life, the nature of consciousness, or perhaps even the very existence of matter in the universe.

Now everybody—

12.6 Further Reading

The following is an excellent English translation of Kepler's little book on snowflakes, accompanied by insightful short essays on Kepler's philosophy of science and contributions to crystallography:

Kepler, J., *The Six-Cornered Snowflake*, translation and reprint, Oxford University Press, 1966.

On snowflakes in general, see

Bentley, W.A., and Humphreys, W.J., *Snow Crystals*, reprint of 1931 McGraw-Hill by Dover Press, 1963;
Komarechka, D., *Sky Crystals*, Don Komarechka Photography, 2013;

and the following two Web sites:

http://skycrystals.ca/snowflake-gallery (viewed August 2016);
http://www.snowcrystals.com (viewed August 2016).

On the physics of snowflake formation, see

Libbrecht, K., "The Physics of snow crystals", *Rep. Prog. Phys.*, **68**, 855–895 (2005).

If you happen to be curious about lichens, I found the Wikipedia page on the topic quite informative:

http://en.wikipedia.org/wiki/Lichen (viewed April 2015).

On solar flares, see the web pages of the SoHO and SDO (Solar Dynamics Observatory) space missions:

http://sohowww.nascom.nasa.gov/ (viewed April 2015);
http://sdo.gsfc.nasa.gov/ (viewed April 2015).

For a detailed presentation of SOC as an explanatory framework for energy release in various astrophysical systems, see

Aschwanden, M., *Self-Organized Criticality in Astrophysics*, Springer, 2011.

Many authors have written on emergence as the key to complexity. On this general topic I always much appreciated the writings of John Holland. If you feel up to it, try

Holland, J.H., *Emergence: From Chaos to Order*, Addison-Wesley, 1998;

and/or his book *Hidden Order*, listed in the bibliography to chapter 2. On science as a way of knowing, see the aptly entitled

Moore, J.A., *Science as a Way of Knowing*, Harvard University Press, 1993.

Finally, should you ever decide to try hiking, canyoneering, or mountain biking (or even unicycling) in the Moab area, keep an eye out for those spiraling crustose lichens!

A

BASIC ELEMENTS OF THE PYTHON
PROGRAMMING LANGUAGE

This appendix is *not* meant to be a comprehensive introduction to the Python programming language. It aims only at presenting—and sometimes providing additional explanations regarding—the use (and possible misuse) of the basic elements of Python on which the codes presented throughout this book are built. The developers of Python and assorted Python libraries have done a pretty superb job of providing online documentation, URLs for which are given at the end of this appendix. Also, never hesitate to google a Python query, you are very likely to get the answer you need (and then some).

With a few exceptions, only syntax elements common to most computing languages are used throughout this book, to ease translation for those wanting to work with a computing language other than Python; going all-out Python could have made many coding constructs more elegant and compact, and run faster, but also made them harder to decipher for non-Python-savvy users. This being said, I found a few Python-specific constructs so useful that I ended up using them; in all cases, their functionality is explained in what follows, and alternative code fragments omitting their use are also provided.

Raw Python is actually a pretty minimal language for the purpose of numerical computation, but these limitations are readily bypassed by the use of various Python libraries. All Python codes provided in the chapters of this book use functions from the NumPy library. Python libraries are still rapidly evolving, but at this point in time, NumPy is a standard. I generally steered clear of high-level functions for scientific computation, to facilitate portability to other computing languages. If you think you need those, look into the SciPy library.

A.1 Code Structure

Python is really a scripting language, so that Python source codes, which should really be called "scripts," are not "compiled" but rather "interpreted." The distinction may be profound from the computer science point of view, but for the user, the most important consequence is that a Python source code is interpreted strictly sequentially, line by line.

Coding certainly carries one's style, and there is definitely satisfaction to be had in writing elegant source code. Style may be personal, but some standard, good programming habits should also be adopted. For example, judicious definition of functions as program subunits is very good programming practice, as it often helps to highlight the overall logic of the code, and favors code modularity. The so-called PEP8 style guide for Python code (see the URL at the end of this appendix) offers useful guidelines on nearly all aspects of Python programming.

A typical self-contained Python source code, such as those presented throughout this book, is structured like this:

```
 1 # ONE OR MORE COMMENT LINES EXPLAINING BRIEFLY WHAT THIS CODE DOES
 2 import numpy as np              # import (and rename) NumPy library
 3 import matplotlib.pyplot as plt # import (and rename) matplotlib library
 4 ...
 5 #-------------------------------------------------------------------
 6 PI=3.1415926536                 # define constants
 7 ...
 8 #-------------------------------------------------------------------
 9 # THIS IS A USER-DEFINED FUNCTION
10 def my_first_function(x,y):
11     ...                          # Python instructions calculating z
12     return z
13 # END FIRST FUNCTION
14
15 # THIS IS ANOTHER USER-DEFINED FUNCTION
16 def my_second_function(z):
17     ...                          # Python instructions calculating s
18     return s
19 # END SECOND FUNCTION
20 #-------------------------------------------------------------------
21 # MAIN PROGRAM
22 ...                              # assorted Python instructions,
23 z=my_first_function(x1,y1)       # including calls to my_first_function
24 # END
```

For a more complete working example see the hodgepodge code listing in figure 11.4. A few things are worth nothing here:

1. You'll have guessed already that # is the comment character in Python. Adding descriptive comments to lines of code is always a good idea. Leaving a line blank has no impact on code execution, but can improve readability, which is also a good idea. In codes of significant length, I like to separate functions from constant definitions, programs, etc., by a comment line of ---, but that's just me.

2. It is often practical to rename libraries upon import; here, for example, the NumPy library is internally renamed np, so that invoking the NumPy function array() can now be done as np.array() instead of numpy.array(). The advantage of such renaming is

perhaps more apparent when invoking functions from a library with a longer name, for example matplotlib.pyplot.

3. The set of instructions associated with a function is identified *only* by being indented to the right with respect to the def header. I highly recommend adding a comment line, as in the above example, to explicitly flag the end of instructions lines pertaining to a function.

4. User-defined Python functions can return more than one value via their `return` statement; see, for example, the `return fx,fy` instruction in the `force` function in figure 10.3.

5. In Python user-defined functions, the `return` statement is optional; functions can operate internally on their arguments, which are then modified upon terminating the function.

6. If you opt to lump everything into a single source file, function definitions must appear *before* being invoked the first time. Otherwise source code for functions must be run through Python prior to running the main program. Consequently, in the above global code structure example, the second user-defined function could call the first, but not the other way around.

7. Any variable declared and initialized prior to the definition of functions and the main program, such as the constant `PI` in the above code fragment, is *global* and as such can be used (but not modified) by any program subunit that follows.

8. Unlike functions, the main program need not be assigned a name. Nostalgic C programmers (like me), desperately longing for `int main(void)`, can include a comment line with `MAIN`, as in the above example.

Python's NumPy library includes the usual set of predefined mathematical functions such as `sin()`, `cos()`, `log()`, and `sqrt()`. If you need it, it most likely exists within Python. Typing the first thing that comes to mind will often get you what you want, otherwise simply fall back on the online documentation.

A.2 Variables and Arrays

Python supports the usual variable types: float, integer, character, Boolean, etc. Variable types need not be explicitly defined, i.e., in interpreting the instruction a=1 Python will assign integer status to the variable a, but would assign floating-point status if the instruction had read a=1.0.

In Python, variable and function names can be longer than you might ever want them to be. Lower- and uppercase characters are distinct, i.e., variables named aa and AA are not the same objects. As in all computing languages, Python reserves certain character strings as keywords for its own internal purposes. You won't need much Python programming experience to figure out that naming a variable `for`, `if`, `else`, `def`, `return`, etc. is probably not a good idea. Some reserved Python keywords are not as intuitively obvious; once upon a time I got into trouble naming a variable `del`, the Python keyword used to delete an element from a list, and the error message I received was not exactly transparent. Other nonintuitive keyword names, all to be avoided as variable names, include `break`, `class`, `in`, `is`, `lambda`, `nonlocal`, `pass`, `try`, and `yield`. To this blacklist must be added the character strings identifying Python's predefined functions such as `min`, `max`, and `range`.

Raw Python supports lists, but the fixed length arrays commonly used in numerical computation are created through specific functions in the NumPy library. Only three are used in this book, and what they actually do depends on the argument provided. For example,

1. `dx=np.array([-1,0,1,0])` creates a 1-D array of length 4 named `dx`, containing the four integer values $-1, 0, 1, 0$;

2. `grid=np.zeros([N,M])` creates a 2-D array named `grid` of size $N \times M$, i.e., of length N in the first dimension (rows/vertical) and M in the second (columns/horizontal) and fills it with the (float) value zero—very useful for initialization (the default variable type is float);

3. `status=np.ones(M, dtype='int')` creates a 1-D array of length M named `status`, and fills it with the integer value 1—also very useful for initialization.

Arrays can also be defined implicitly through mathematical operations, or the `return` of a function. For example, if `a` has already been defined as a 1-D array of length N, the instruction `b=a` will create an second array `b` of length N and fill it with the corresponding elements of `a`. Likewise, in the code fragment presented above, if `z` returned by the first function is a 2-D array of size $N \times N$, the instruction

```
q=my_first_function(x,y)
```

will create a 2-D $N \times N$ array named `q` and fill it with the elements of the local array `z` calculated internally within that function.

Individual array elements are accessed through their *index*, giving their position within the array. Python numbers elements of an array of length N from 0 to $N-1$, so that `a[1]` accesses the *second* element of array `a`, `a[N-1]` the last, and `a[N]` will blow you out of array bounds. You get used to it eventually.

A.3 Operators

Python includes all the basic arithmetical operators, using the usual keyboard symbols +, -, *, / for addition, subtraction, multiplication, and division, respectively. Note that, unlike in many computing languages, in Python explicit integer division such as 7/2 will return 3.5, rather than 3; in other words, implicit conversion to real type will take place. If you really want to divide two integers and get a truncated integer result, you must use Python's integer division operator //; for example, 7//2 will return the integer 3.

Python allows very flexible use of the value assignment operator =, for example the one-line instruction a,b,c=0,1.,0 sets a=0 (integer), b=1.0 (float), and c=0 (integer). Powers use the old FORTRAN syntax **, i.e., a**2 is the same as a*a, and fractional exponents are allowed, so that, for example, a**(1/3) returns the cube root of a.

Python also includes many other arithmetical operators, some quite useful, for instance the additive/multiplicative increment/decrement operators +=, -=, *=, and /=, where

a+=b is equivalent to a=a+b,   a-=b is equivalent to a=a-b,
a*=b is equivalent to a=a*b,   a/=b is equivalent to a=a/b.

Another very useful operator is the modulus %, such that a%b returns the remainder of the division of (positive) integer a by (positive) integer b. This is particularly useful to enforce periodicity to random walks on lattices (see, e.g., the ant code listed in figure 2.10). Consider the instruction

```
ix=(N+ix) % N
```

As long as $0 \leq ix < N$, then $N \leq N + ix < 2N$, so that the above instruction will leave the value of ix unchanged; but if $ix < 0$, then this instruction will add N to ix; and will subtract N if $ix \geq N$.

Under Python's NumPy module, arithmetical operators can also act on arrays. For example, if a and b are two 1-D arrays of length N, the instruction

```
c=a+b
```

creates an array c also of length N, and sets its elements equal to the pairwise sum of the elements of a and b. This is equivalent to the following instructions:

```
c=np.zeros(N)
for i in range(0,N):
    c[i]=a[i]+b[i]
```

This works only if the arrays have the same dimensions and lengths, otherwise Python will return a run-time error. However, one useful Python/NumPy-legal possibility, used in this book, is to add a scalar to every element of an array. For example, in the earthquake code of figure 8.3, lattice driving takes place by adding the same scalar increment delta_f to every element of the 2-D array force through the instruction

```
force[:,:]+=delta_f
```

where the : symbol signifies "all elements of this array dimension."[1]

A.4 Loop Constructs

Python supports the usual two basic loop constructs: fixed-length (**for**) loops and conditional (**while**) loops. The basic syntax for a

[1]The shorter instruction force+=delta_f would be Python-legal as well, and achieve the same result. I find this to be potentially confusing when reading the code, and so such syntax is avoided everywhere in this book.

fixed-length loop is the following:

```
for i in range(0,N):
    ...
```

where ... stands for one or more lines of syntactically correct Python instructions. This loop would repeat N times, with the loop index variable i running from 0 to $N - 1$; i.e., *not* from 0 to N, as the colloquial meaning of "range" would normally suggest. A third, optional parameter can be provided to range, controlling the size of the increment for the loop control variable; writing the above as for i in range(0,N,2) would run the loop with values $i = 0, 2, 4, 6, 8, \ldots, N - 1$ (or $N - 2$ if N is even). It takes a little while to get used to this convention, but it works, and has at least the merit of being compatible with array indexing, in which elements of an array of length N are also indexed from 0 to $N - 1$.

The fixed-length loop structure just described runs over a preset number of iterations, determined by the two arguments given to range(). In some situations it might not be possible to determine a priori the number of iterations required by a loop. For example, in the DLA simulations of chapter 3, the temporal iteration only needs to run until all particles are stuck, or, in the epidemic simulation of chapter 9, until the number of infected individuals has fallen to zero. The appropriate temporal loop construct in such a case would be, as in the epidemic code of figure 9.1,

```
max_iter=100000          # maximum number of iterations
...
iterate=0                # iteration counter
while (n_infect>0) and (iterate<max_iter): # temporal loop
    ...                  # line(s) of Python instructions
    iterate+=1           # increment iteration counter
# end of temporal loop
```

Note that the loop control condition includes a safety test ensuring that the loop cannot run forever, if some algorithmic design flaw or coding mistake were to cause n_infect to never fall to zero.[2]

Something equivalent to conditional loops can also be constructed using the break instruction, which prematurely exits an ongoing loop and picks up execution with the first instruction following the end of the loop. As a specific example, the conditional loop above could be written instead as

```
max_iter=100000            # maximum number of iterations
...
for iterate in range(0,max_iter): # temporal loop
    ...                     # line(s) instructions
    if n_infect == 0: break # break out of loop prematurely
# end of temporal loop
```

Such use of the break statement to build conditional loops is often not considered good programming style, but it can be useful in some circumstances.

A particularly objectionable (IMHO) feature of loop syntax in Python is that the block of instructions acted upon by the loop is identified *only* by being indented with respect to the loop instruction, which means, for example, that

```
for i in range(0,N):
    a=i+1
    print("a= {}.".format(a))
```

will *not* produce the same output as

```
for i in range(0,N):
    a=i+1
print("a= {}.".format(a))
```

---

[2]I highly recommend developing this to a reflex when coding **while** loops.

In the first case, the value of a would be printed to screen at every iteration of the loop, but in the second case only the last value would, after exiting the loop. If the loop controls only a few lines of instructions, this indentation-based loop-delimiting syntax is tolerable; but for loops containing many instructions, or other nested loops, or conditional blocks of instructions, the code logic can become harder to follow. As a compromise, in many of the codes listed in this book, I have added a comment line to explicitly mark the end of long instruction blocks associated with loops or conditional statements, as exemplified in the **while** loop example above.

Note finally that if a loop controls a single line of instruction, all of this can be written on the same line, as in line 6 in the box-count code of figure 3.10:

```
while (2**n_scales<n) and (n_scales<100): n_scales+=1
```

Like many modern programming languages, Python also supports a form of implicit loop defined using the symbol :, used to access a subset of contiguous array elements. For example, if A is an array of dimension 1 and length N, writing A[i1:i2] accesses elements i1 to i2-1 of the array. This may seem straightforward, but where it becomes potentially confusing is in a statement like A[0:10], which accesses the 10 elements indexed from 0 to 9 of array A. To make things worse, *this convention is different in many other computing languages*, where the equivalent of the Python A[0:10] syntax would mean "access array elements indexed 0 through 10 inclusively," now for a total of 11 elements. Seasoned MATLAB and IDL programmers, beware!

This being said, the syntactic shortcuts allowed by the use of : are just too useful to skip. Consider, for example, the following instruction in the source code for cluster tagging listed in figure 4.3:

```
map_cluster[1:N+1,1:N+1]=lattice[:,:]
```

In a single instruction line, this copies the $N \times N$ array `lattice` into the larger $(N + 2) \times (N + 2)$ array `map_cluster`, leaving all edge values (ghost nodes) of `map_cluster` at their initialized zero value. This instruction is thus equivalent to the double loop construct

```
for i in range(0,N):
    for j in range(0,N):
        map_cluster[i+1,j+1]=lattice[i,j]
```

A.5 Conditional Constructs

Python includes all the usual **if...** and **if...else...** conditional constructs, with logical conditions expressed in terms of the (self-explanatory) operators <, >, <=, >=, as well as the somewhat less self-explanatory == and != for "equal to" and "not equal to," respectively. Conditional statements constructed in this manner can be combined using the usual and and or logical operators. Two examples should suffice to illustrate the concept, the first taken from the epidemic code of figure 9.1. The block of instructions following

```
if (infect[k] == 0) and (k != j):
```

is executed provided both conditions within parentheses are satisfied (i.e., evaluate to Boolean TRUE).[3] The second example is taken from the earthquake code of figure 8.3:

```
if toppling[iterate] > 0:
    force+=move
else:
    force[:,:]+=delta_f
```

---

[3]The parentheses "(...)" are optional but I highly recommend their use in such compound conditional expressions.

Here the 2-D lattice `force` is updated by addition of the 2-D array `move` if at least one toppling has occurred (first block of instructions), otherwise the scalar increment `delta_f` is added at every node of the lattice (second block of instructions). As with loop constructs, the blocks of instructions controlled by the conditions are *only* delimited by being indented to the right (I really hate this!), but a single condition-controlled instruction can be included after the colon as a single instruction.

Python does not include a straightforward **case** (or **switch**) construct; these must be built using sequential **if** or nested **if. . . else** statements, or the Python contracted version **elif**. See the lattice update rules in the hodgepodge code listed in figure 11.4 for a specific example.

One type of Python-specific conditional instruction is so useful that I opted to make use of it in some of the codes listed in this book. For example, in figure 4.3,

```
if iic in map_cluster[jj+dx[:]],kk+dy[:]]:
    ... # instruction(s) subject to conditional execution
```

This searches for the presence of the value `iic` in any one of the nearest neighbors of node `[jj,kk]`, as defined by the four element pairs stored in the template arrays `dx` and `dy`. This single instruction is here equivalent to the construct

```
ifound=0
for i in range(0,4):
    if map_cluster[jj+dx[i],kk+dy[i]] == iic:
        ifound=1
if ifound == 1:
    ... # instruction(s) subject to conditional execution
```

A.6 Input/Output and Graphics

Python includes the usual set of functions for writing or reading to files, printing to screen, or reading keyboard input. The only one used in the codes listed throughout this book is the basic "print to screen" function `print`; see, for example, line 43 in the DLA code of figure 3.1 for a specific example. "Pretty printing," with full control over format, is of course possible. See the Python documentation for more on all this I/O stuff.

The output of most simulations described throughout this book is usually best displayed as pixelated images (for any simulation defined over a lattice), or even better, animation of such images. The Python library matplotlib contains many user-friendly graphical functions that do exactly this. See the URL provided below. Most codes included in this book use only very basic plotting instructions, all using matplotlib.

Some simulations, for example the forest-fire model of chapter 6, the flocking simulations of chapter 10, or the hodge-podge spiral simulations of chapter 11, are most definitely best appreciated as animations; unfortunately, there is as yet no really easy way to do this in Python. The closest is the mat-plotlib function `funcAnimation`, but it requires encapsulating the simulation time steps as a function to be called by the animation function. I opted to leave this out of the codes provided throughout this book, but it is definitely worth the effort. Note also that precomputed animation files for some of these simulations are available via the Princeton University Press website at

http://press.princeton.edu/titles/11053.html.

A.7 Further Reading

The official home page of the Python programming language is

www.python.org.

It gives (free) access to the software, provides download and installation instructions, user's guides, and beginner's tutorials, as well as many code examples. Their Python tutorial contains pretty much everything you need to know (and a lot more) to work through this book:

www.pythonprogramminglanguage.com.

The PEP8 style guide is also available here:

www.python.org/dev/peps/pep-0008.

Two other excellent Python resources are

Langtangen, H.P., *A Primer on Scientific Programming with Python*,
4th ed., Springer (2014);
Swaroop, C.H., *A Byte of Python*, www.swaroopch.com/notes/python/,

as well as the Python tutorial from Code Academy:

www.codeacademy.com.

For programming beginners, the Python tutor is excellent:

www.pythontutor.com.

On the NumPy and matplotlib Python libraries, see

www.numpy.org,
www.matplotlib.org.

At this writing, the easiest way to get started downloading and installing these (and other) Python libraries is through either one of the following open platforms:

www.scipy.org,
www.continuum.io.

No point procrastinating, start downloading now!

B

PROBABILITY DENSITY FUNCTIONS

Probability density functions (PDFs) measure the probability of finding a measurement in some specified interval of possible values for the measured quantities. As analysis and interpretative tools, they are used repeatedly in this book, and a basic understanding of their construction and interpretation is essential. This is the aim of this appendix. Appendix B.1 introduces the idea at the pre-calculus level through a simple example, while the following sections require a working knowledge of the calculation of derivatives and integrals of functions of a single variable.

B.1 A Simple Example

The following numbers are the grades (in percent) obtained a few years ago by my cohort of $N = 83$ undergraduate students on the midterm exam of my Introduction to Computational Physics class:

[46, 84, 70, 66, 41, 82, 69, 59, 28, 81, 88, 82, 83, 33, 27, 51, 62, 72, 87, 55, 66, 68, 55, 86, 75, 74, 56, 81, 60, 44, 84, 86, 75, 34, 96, 45, 57, 79, 81, 52, 24, 38, 74, 89, 68, 85, 85, 45, 62, 96, 45, 40, 48, 90, 46, 57, 33, 71, 67, 82, 94, 43, 16, 88, 46, 91, 82, 55, 71, 86, 77, 63, 81, 78, 59, 84, 100, 69, 92, 69, 44, 64, 88].

Let g_k represent the grade obtained by the kth student. The class average, $\langle g \rangle$, for this exam is simply given by the sum of all grades divided by the class size:

$$\langle g \rangle = \frac{1}{N} \sum_{k=1}^{N} g_k, \qquad (B.1)$$

which for the above data is 66.3. To what degree is this number really representative of students' grades? This information can be obtained by constructing the PDF of the grades.

For such a discrete data set, an approximation to the PDF can be built by constructing a *histogram*. This consists in dividing the allowed range of the measured variable—here grades between 0% and 100%—into contiguous bins, each spanning a range of grades, and then counting how many data points fall into each bin, for example counting how many students have a grade between 60% and 64.99%, or between 65% and 69.99%. This defines a discrete function

$$h_m(g; b), \quad m = 1, \ldots, M, \qquad (B.2)$$

where b is the bin size and h_m is the count in the mth bin. The number of bins is simply $M = 100/b$, i.e., the numerical extent of the data, here 100, divided by the bin size.

Figure B.1 shows histograms of the above data, for bin sizes of 2, 5, and 10, and with the class average indicated by the vertical dashed-line segment.[1] The heights of the histogram bins obviously depend on the choice of bin size; for a data set of a given length, the wider the bins the higher the corresponding counts, and some bins can of course remain empty. No matter the bin size, in all cases the sum of counts in all the bins is always equal to the class size $N = 83$; in other words, the histogram can be

---

[1] Since a histogram is, fundamentally, a discrete function of the measurement variable, it is customary to plot it in so-called histogram mode, i.e., as a piecewise-constant function, varying discontinuously at bin boundaries.

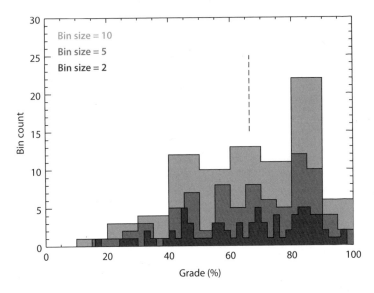

Figure B.1. Histograms for my midterm exam data set, for bin sizes of 2 (dark gray), 5 (gray), and 10 (light gray). The vertical dashed-line segment is the class average computed using equation (B.1).

normalized:

$$\frac{1}{N} \sum_{m=1}^{M} h_m = 1. \qquad (B.3)$$

Knowing h_m also allows an alternative procedure to compute the class average:

$$\langle g \rangle = \frac{1}{N} \sum_{m=1}^{M} h_m \times b_m, \qquad (B.4)$$

where b_m is the grade value at the center of bin m.

Python's NumPy library includes a function, named numpy.histogram(), that accepts as input an array of values, such as my midterm exam grades above, and returns an array containing histogram bin counts (10 equidistant bins by default). It is also possible to set the number of bins, their sizes and ranges through the function's argument list; see the SciPy/NumPy documentation (URLs provided at the end of appendix A).

To turn the counts of figure B.1 into a probability, we need to divide them by the class size $N = 83$, and to turn them into a probability *density* we also need to divide by the bin size. This last step is required so that the quantity $h_m \times b$ measures the probability p of finding, in the grade data set, a grade falling between the bounding values of each corresponding bin. This is the very definition of a probability density. Figure B.2 shows the result of this procedure for the histogram of bin size $b = 5$ from figure B.1, and defines the *discrete* PDF for this data set. Its detailed shape is obviously influenced by the chosen bin size, and some of the finer structure also reflects specificities of the underlying data set; figure B.2 would not be identical if I had used midterm exam data from a different year, even though my average grade for the midterm exam in this course always hovered around 65%. The finite size of the data set also guarantees that the PDF will not be smooth, and this would remain the case even if I opted to minimize my grading time by assigning purely random grades in the allowed range (more on this point in appendix C.2 below). But since I did not engage in such accelerated grading practices, the PDF of figure B.2 does capture something about students' performance (as well as my grading performance, presumably). Note how the most probable grade, i.e., the bin with the highest probability density (0.0289 for the bin 80–85%), is not the one spanning the average grade (bin 65–70%, with a probability density of 0.0193), reflecting the fact that this distribution is asymmetric about its mean value. I observed such an asymmetry almost every year I taught this course; it was something real.[2]

---

[2]In fact, my grade PDFs often could be reasonably well fit by a combination of two Gaussians, one very broad and centered around 60%, the other much

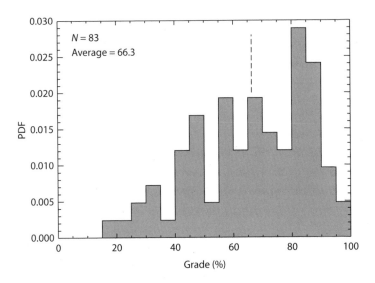

Figure B.2. Discrete PDF constructed from the gray histogram of figure B.1 (bin size = 5). The shapes are, of course, identical on both plots; only the vertical scale is altered, from raw counts to probability density.

B.2 Continuous PDFs

In the limit of very large data sets and infinitesimally small bin size (in the calculus sense), the PDF can be considered a smooth, continuous function $f(x)$, where the quantity $f(x)\,dx$ measures the probability of finding a measurement in the interval $[x, x + dx]$. The equivalent of (B.3) becomes the normalization

---

narrower and centered around 80%. These could be traced pretty directly to two distinct groups of students in the class: physics major students, taking the course in their first semester, and physics-math double-major students, taking the class in the third semester. The latter group dominated the PDF peak around 85%, indicating rather unambiguously that my midterm exams favored them unfairly. Seeing this pattern repeat itself year after year, I eliminated the midterm exam from the course evaluation.

constraint

$$\int f(x)\, dx = 1, \tag{B.5}$$

and the equivalent of (B.4) is then

$$\langle x \rangle = \int f(x)x\, dx. \tag{B.6}$$

In both cases the integral must cover the full range of the variable x.

B.3 Some Mathematical Properties of Power-Law PDFs

Probability density functions of event sizes taking the form of power laws are common in many of the natural and simulated systems considered in this book, so they deserve a bit more attention. Such a power-law PDF is written as

$$f(x) = f_0 x^{-\alpha}, \quad \alpha > 0, \quad x \in [x_0, x_M], \tag{B.7}$$

where the constant f_0 is set by the normalization constraint (B.5). Without loss of generality we can set the normalization interval of the PDF to the range $[1, \infty]$, so that the substitution of (B.7) into (B.5) yields

$$f_0 = \lim_{x_M \to \infty} \frac{\alpha - 1}{1 - (x_M)^{1-\alpha}}. \tag{B.8}$$

Evidently, the normalization is only possible provided $\alpha > 1$, otherwise $f_0 \to \infty$ in the limit $x_M \to \infty$. For normalizable PDFs we have

$$f_0 = \alpha - 1, \quad \alpha > 1, \tag{B.9}$$

in the $x_M \to \infty$ limit. Now consider a situation where the variable x is extracted from measurements spanning the range $[x_0, x_M]$, as described by a normalizable power-law PDF ($\alpha > 1$). In such a situation the normalization constant becomes $f_0 = (\alpha - 1)/(x_0^{1-\alpha} - x_M^{1-\alpha})$, and the mean value of the variable calculated via equation (B.6) is

$$\langle x \rangle = \frac{f_0}{\alpha - 2}(x_0^{2-\alpha} - x_M^{2-\alpha}), \quad (\alpha \neq 2). \qquad (B.10)$$

In many cases the PDF spans many orders of magnitude in the variable x, i.e., $x_0 \ll x_M$. We can then distinguish two regimes:

- $1 < \alpha < 2$. This implies $2 - \alpha > 0$, so that the term involving x_M dominates in (B.10). We then have

$$\langle x \rangle = \frac{f_0}{2 - \alpha} x_M^{2-\alpha}. \qquad (B.11)$$

- $\alpha > 2$. This implies $2 - \alpha < 0$, so that the term involving x_0 dominates (B.10), in which case,

$$\langle x \rangle = \frac{f_0}{\alpha - 2} x_0^{2-\alpha}. \qquad (B.12)$$

The special case $\alpha = 2$ is "left as an exercise," as we like to say in the business.

These two distinct regimes, as delineated by the value of the power-law index α, have important consequences when constructing a PDF from a finite set of individual measurements. Note in particular that, for $\alpha < 2$, the average event size is determined by the largest measured event. These being rare if the PDF is a power law, computing the mean event size from an experimental or numerical data set containing too few events could lead to a gross underestimate of the mean value. This is no longer the case if $\alpha > 2$, since the mean value is then dominated by the smaller, more frequent events, which will be well represented even in a (relatively) small data set. If $\alpha < 1$, the mean value cannot even be mathematically defined.

B.4 Cumulative PDFs

Sometimes, observational data are represented through a *cumulative* PDF $f(> x)$, such that $f(> x) \, dx$ measures the probability of finding a measured value larger than x. Our encounter with the Gutenberg–Richter law in chapter 8 offered one example. If x is distributed as a power law, then we have

$$f(> x) = \int_x^\infty f_0 x^{-\alpha}\, dx = x^{1-\alpha}, \qquad (B.13)$$

where the second equality holds only if the distribution can be normalized, requiring $\alpha > 1$, so that $f_0 = \alpha - 1$. In such a situation, the cumulative PDF is also a power law, with an index differing by unity as compared to the index of the usual noncumulative PDF.

B.5 PDFs with Logarithmic Bin Sizes

If the PDF of a measured variable takes the form of a power law, the tail of a PDF constructed from measurements will contain very few events, and so will be very "noisy," making it difficult to reliably infer the numerical value of the power-law exponent α. This is illustrated in figure B.3, for PDFs constructed from a set of $N = 300$ data points extracted from a power-law distribution with index $\alpha = 1.75$. In panel A, the PDF uses a bin size $b = 10$ and is plotted using the usual linear axes. Because the PDF falls off very rapidly with increasing x, here most points end up concentrated in the first bin ($0 \leq b < 10$). When replotting the same data using logarithmic axes, as in panel B, bins for $x > 100$ either contain only one point, or none at all. Fitting a straight line to this PDF looks like a pretty risky proposition. Turning to the cumulative version of the PDF, as shown in figure B.3C, improves the situation somewhat, in that the middle of this distribution could conceivably be fit with a straight line to yield the exponent $\alpha - 1$ (= 0.75 here, and indicated by the dashed line segment). However, choosing the start and end points of the fitting regions will be very tricky unless the PDF spans many orders of magnitude in the measured variable.

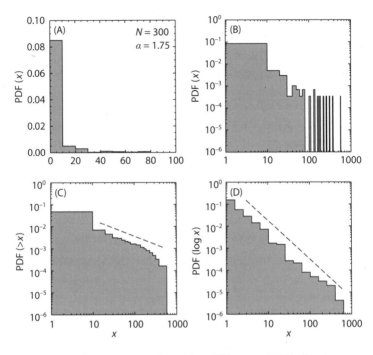

Figure B.3. Plotting can make such a difference. PDFs for the same data are plotted in these four panels, built from a set of 300 synthetic data points extracted from a power-law distribution with index $\alpha = 1.75$. Panel A plots the PDF on a standard plot with linear horizontal and vertical axes, with a bin size of 10 used in constructing the PDF. Panel B plots the exact same data, now using logarithmic axes, and panel C plots the cumulative PDF of the same data, again plotted on logarithmic axes. The resolutely reckless could consider fitting a straight line through the middle decade of the distribution. Panel D plots a PDF of the same data still, now constructed using a logarithmic bin size of $\log(b) = 0.2$ and again plotted on logarithmic axes. A straight-line fit can now be contemplated with some measure of confidence. The dashed lines indicate the true slope of the distribution from which the $N = 300$ data were drawn (minus unity for the cumulative PDF in panel C, as per equation (B.13)).

One way around this difficulty is to introduce bin sizes that increase with x. A particularly simple way to achieve this is to construct the histogram function of the *logarithm* of the measurement variable:

$$y = \log(x) \rightarrow dy = \frac{dx}{x}, \tag{B.14}$$

and then use a bin size that is constant in y. Whatever the variable we use to construct the PDF, the normalization constraints (B.5) always hold, so that

$$\int f(x)\,dx = \int f(y)\,dy = \int \frac{f(y)}{x}\,dx, \tag{B.15}$$

using the second equality resulting from equation (B.14). We thus conclude that

$$f(x) = \frac{f(\log x)}{x}. \tag{B.16}$$

In other words, we are correcting the counts (and associated probabilities) by accounting for the fact that the bin size increases linearly with x. In this way, even if the PDF is constructed as $f(\log x)$ with logarithmically constant bins, for a power-law PDF, we find that plotting $f(\log x)/x$ versus x using logarithmic axes will still yield a straight line, with slope corresponding to the index α of the underlying PDF $\propto x^{-\alpha}$ for the original measurement variable. Figure B.3D shows the result of this procedure, using the same underlying data as in panels A and B. A linear fit (on the log-log plot representation) can now be envisioned with some measure of optimism.

From figures 4.8 to 8.7, all power-law PDFs encountered in this book are constructed and plotted in this manner. When the measured variable spans many orders of magnitude and N is very large, a straight linear least-squares fit can then return a reasonably accurate estimate of the power-law index α. If either of these conditions is not satisfied however, the inferred value of α may deviate significantly from the underlying "true" value; fortunately, it is (relatively) easy to do better.

B.6 Better Fits to Power-Law PDFs

Compared to the dashed line indicating the true logarithmic slope of the distribution from which the synthetic data were drawn, the PDF in figure B.3 looks pretty good in a "chi-by-eye" sense. Yet a formal linear least-squares fit with equal weight assigned to each bin yields $\alpha = 1.697 \pm 0.051$, somewhat lower than the true underlying value $\alpha = 1.75$. Maybe that's "good enough," or maybe not. Remember the dangers of earthquake prediction encountered at the end of chapter 8.

A proper statistical approach to this fitting problem would be to infer the index for the power-law distribution that has the highest likelihood of having generated the N measured data. If x is a continuous variable and its PDF is normalizable (i.e., $\alpha > 1$), this maximum likelihood estimator for α is

$$\alpha = 1 + N \left(\sum_{n=1}^{N} \ln \frac{x_n}{x_{min}} \right)^{-1}, \qquad (B.17)$$

where x_{min} is the lower bound of the range within which power-law behavior holds, as determined empirically from the data or on theoretical grounds. The associated standard error (σ) on α is given by

$$\sigma = \frac{\alpha - 1}{\sqrt{N}}. \qquad (B.18)$$

For the synthetic data of figure B.3 and with $x_{min} = 1$, the above expressions yield $\alpha = 1.744 \pm 0.043$, much closer to the target 1.75 than a linear least-squares fit to the log-log plot of logarithmically binned data in panel D.

In working with real data (including measurements from lattice-based simulations), two difficulties must be dealt with. The first is the choice of x_{min} in equation (B.17). Consider, for example, the PDF of percolation cluster sizes plotted in figure 4.8. As argued in section 4.5, the self-similar fractal structure cannot be expected to extend down to clusters of size 1. Looking at figure 4.8, picking $x_{min} = 10$ might be a reasonable "chi-by-eye" choice; a higher value would obviously be safer, but this would also mean running the risk of throwing away more potentially useful data, a problem that can become ever more acute the steeper the power law. Here again, statistically sound approaches are available to pick a proper x_{min} (see the bibliography at the end of this appendix).

A second potential difficulty arises from the fact that the upper end of the distribution of percolation cluster sizes is likely to be affected to some extent by the finite size of the lattice, in the sense that the size of the PDF cannot be expected to drop instantaneously to zero at exactly the largest percolation cluster size $p_c N^2/2$. Different strategies exist to augment such power laws with an upper cutoff, finite-size scaling function. Its defining parameters must then be fit simultaneously with those of the power law, which usually results in a nonlinear fitting problem even if carried out in log-log space. See, for example, the book by Christensen and Moloney cited in the bibliography of chapter 4.

This second difficulty usually does not arise when working with real-world data, for which a hard upper limit is seldom expected. For example, the largest earthquake ever measured, the May 22, 1960 earthquake in Chile, scored 9.5 on the Richter magnitude scale; yet nothing in plate tectonics precludes, in principle, more energetic earthquakes; they simply have not occurred since the beginning of the earthquake-magnitude record. Likewise, the solar flare of figure 12.4 is in all likelihood the largest observed during the space era, but the observations of "superflares," up to 10^4 times more energetic, on stars other than the Sun confirms that nothing close to the upper limit has yet been observed on the Sun—and, as with earthquakes, we can only hope it stays that way.

B.7 Further Reading

Most statistics textbooks discuss PDFs at some level. See, for example,

James, F., *Statistical Methods in Experimental Physics*, 2nd ed., World
    Scientific (2006);

Roe, B.P., *Probability and Statistics in Experimental Physics*, Springer
    (1992).

On the inference of power-law behavior in experimental data, see

Clauset, A., Shalizi, C.R., and Newman, M.E.J., "Power-law distribu-
    tions in empirical data," *SIAM Review*, **51**(4), 661–703.

C

RANDOM NUMBERS AND WALKS

C.1 Random and Pseudo-Random Numbers

A sequence of numbers is said to be *random* if the numerical value of each member in the set is entirely independent of the numerical value of the other members of the set. Once upon a time, I enlisted my then six-year-old son to roll a standard six-faced die 12 times in a row, twice so; the results were the following two sequences:

$$4 - 2 - 6 - 3 - 4 - 4 - 2 - 6 - 1 - 5 - 2 - 6,$$
$$6 - 6 - 1 - 2 - 3 - 3 - 5 - 6 - 3 - 6 - 2 - 3.$$

You may note that the second sequence does not include a single "4." This is not so surprising as one may think, considering that the probability of *not* rolling a "4" is $1 - 1/6 = 5/6$, so that the probability of not rolling a "4" 12 times in a row, is $(5/6)^{12} = 0.112$. This is small, but certainly not astronomically so (unlike your odds of winning at the lottery, which are). If indeed each throw is entirely independent of the preceding throw, then the odds of obtaining exactly one of these sequences is $(1/6)^{12} = 4.6 \times 10^{-10}$, which is in fact exactly the same as obtaining one of the following two sequences, which most people would judge, incorrectly, to be far less probable:

$$1 - 2 - 3 - 4 - 5 - 6 - 1 - 2 - 3 - 4 - 5 - 6,$$
$$6 - 6 - 6 - 6 - 6 - 6 - 6 - 6 - 6 - 6 - 6 - 6.$$

Now, if you roll a die a very great many times (N, say), then you would expect to roll "1" $N/6$ times, "2" $N/6$ times also, and so on to "6." If you get different numbers, then you should really take a closer look at that die. For an unloaded die, every roll is independent of the others, and every one of the six possible outcomes is equiprobable. In other words, the die is a *generator* of random integers uniformly distributed in the interval [1, 6], and die-throwing is categorized as a stationary, memoryless random process.

How do you achieve the same thing on a computer? At first glance this may appear nonsensical, considering that a computer program is entirely deterministic; on a given architecture, an executable program will always return the same output upon being presented with the same input. This means that a computer program autonomously simulating successive throws of a die will always produce the same sequence. In other words, the result of the nth "throw" will be entirely determined by the state of the computer's memory following the $(n - 1)$th throw, this being true for all throws. Successive throws in the sequence are completely correlated; we could be no further from a memoryless random process.

The way out of this paradox is to accept the fact that successive throws will be perfectly correlated, but design our die-throwing algorithm so that, *over a long sequence of throws*,

- every throw value is equiprobable;
- there is no statistical correlation between successive throws; in other words, a "1" anywhere in the sequence is as likely to be followed by a "1," a "2," a "3," etc.

It is this *statistical uniformity* of the sequence that defines its random status, although the term "pseudo-random" is usually preferred, to distinguish it from truly random sequences, such as die throws, coin flips, or radioactive decay.

C.2 Uniform Random Deviates

Many simulation codes listed in this book require either a random number generator which returns floating-point numbers uniformly distributed in some fixed interval, or integers distributed uniformly in some range $[0, N]$. Python's NumPy library contains such generators (and many others). Generic random number generators exist in most programming languages. The theoretical, arithmetical, statistical, and computational underpinnings of the generation of pseudo-random numbers are rather intricate and would fill many pages, but this would not be particularly useful here. For the purposes of working through this book, all you need to know is that pseudo-random number generators do exist, some are better than others, a few are downright crappy, and by now the truly objectionable among these have gone extinct.

As an example, what follows are two distinct ways to generate the computational equivalent of rolling a six-faced die:

```
import numpy as np
...
roll=np.random.random_integers(1,6)
...
```

Note that here, the upper and lower bounds given are inclusive (*unlike* the `range()` function controlling unconditional loops in Python), so that the above call will return 1, 2, 3, 4, 5, or 6 equiprobably. The other way is

```
import numpy as np
...
roll=np.random.choice([1,2,3,4,5,6])
...
```

Sometimes it is necessary to generate distinct sequences of random numbers, for example when testing different realizations of a stochastic process for the purposes of ensemble averaging; many such instances can be found in this book. The NumPy library includes a function named numpy.random.seed(), which allows the numerical value of the seed to be set for subsequent calls to any one of Python/NumPy's random number generators, by passing a specific integer value as an argument, for example numpy.random.seed(1234).

C.3 Using Random Numbers for Probability Tests

The ability to generate random numbers uniformly distributed in the unit interval allows a simple numerical implementation of probability tests, in simulations involving stochastic rules. In the forest-fire model of chapter 6, for example, a tree on a lattice node can be ignited by lightning with probability p_f. For each such tree, the "decision" to ignite or not can be encapsulated in a one-line conditional statement:

```
if np.random.uniform() < p_f: # lightning strikes (maybe)
```

Since successive draws of the random number r (as produced by np.random.uniform()$\equiv r$) are uniformly distributed in $[0, 1]$, then for $p_f = 10^{-5}$ (say), on average one draw in 10^5 will satisfy $r < p_f$. Consequently, on average one in every 10^5 trees will be ignited by lightning at each temporal iteration. If the lattice is very large and contains a number of trees $\gg p_f^{-1}$, then many trees will be ignited at each iteration; in the opposite situation, ignition events will be separated in time, with the wait time between successive lightning strikes distributed exponentially in the regime $p_f \ll p_g$.

The above procedure effectively draws pseudo-random numbers from a PDF (see appendix B) of the form

$$f(x) = \begin{cases} 1, & 0 < x \leq 1, \\ 0 & \text{otherwise}, \end{cases} \qquad \text{(C.1)}$$

which satisfies the normalization condition (B.5). However, the discrete PDF for a sequence of N pseudo-random numbers, constructed following the procedure described in appendix B.1, will converge to equation (C.1) only in the limit $N \to \infty$. Figure C.1 illustrates this convergence, for N-member sequences of pseudo-random numbers with N increasing from 300 in panel A to 300,000 in D. In all cases, the bin size is $b = 0.05$, so that $M = 20$ bins are required to cover the interval. This PDF being normalized, the expected value of every bin, in the limit $N \to \infty$, is unity. Clearly, fluctuations about this expected value decrease rapidly as N increases. This decrease can be quantified by computing the *root-mean-squared* deviation about the expected value:

$$\sigma = \left(\frac{1}{M} \sum_{m=1}^{M} (b_m - 1)^2 \right)^{1/2}. \qquad \text{(C.2)}$$

The dotted lines in figure C.1 indicate the range $\pm\sigma$ about the expected value of unity. As with any stationary, memoryless, random process, σ varies as $1/\sqrt{N}$.

C.4 Nonuniform Random Deviates

In some situations it can be useful, or even necessary, to produce sequences of pseudo-random numbers extracted from nonuniform probability distributions. Python's NumPy library contains many functions producing various common distributions of random deviates. If you only have access to a function providing uniform random deviates, it is still possible to generate other types

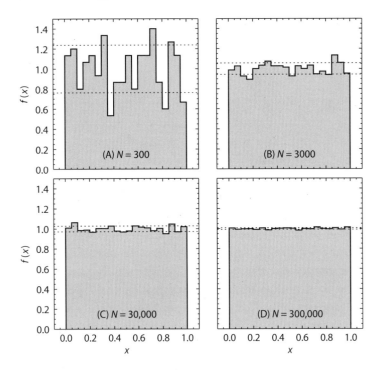

Figure C.1. Probability density functions constructed from sequences of N random numbers uniformly distributed in the interval $[0, 1]$, with N increasing by successive factors of 10 going from panels A through D. The horizontal dotted lines indicate the range $\pm\sigma$ about the expected value $f(x) = 1$.

of distributions, through a technique known as the *transformation method*. What follows states only a few useful results. In all cases, r is a random deviate extracted from a uniform unit distribution, $r \in [0, 1[$, and x is the sought deviate from another distribution.

To get uniform random deviates in the range $[a, b]$, a simple linear rescaling does the job:

$$x = a + (b - a)r, \quad r \in [0, 1], \quad x \in [a, b]. \qquad \text{(C.3)}$$

In this case, the mean of the distribution is $\langle x \rangle = (a + b)/2$. This can also be achieved by the Python/NumPy function call x=numpy.random.uniform(a,b).

An exponential deviate in the range $x \in [0, \infty]$ is given by

$$x = -\lambda \ln r, \quad r \in [0, 1], \quad x \in [0, \infty]. \quad \text{(C.4)}$$

Note that for $r \in [0, 1]$, $\ln(r) < 0$; the minus sign in equation (C.4) is important, don't forget it! Here, the parameter λ sets the scale of the exponential falloff; smaller values of λ give a more steeply peaked distribution, and larger values a flatter exponential distribution. In all cases, the mean value of the distribution, as given by equation (B.6), is $\langle x \rangle = \lambda$, even though the most probable value is zero.

Under Python's NumPy library, the function call x=numpy.random.exponential(scale) produces such exponential deviates, but do note here that scale$\equiv 1/\lambda$ in the above expressions.

For a power-law PDF (see equation (B.7)) normalized to unity in the range $[1, \infty]$, the required transformation is

$$x = r^{1/(1-\alpha)}, \quad r \in [0, 1], \quad x \in [1, \infty], \quad \alpha > 1. \quad \text{(C.5)}$$

The artificial data used to generate the PDFs in figure B.3 were generated in this manner.

The function numpy.random.power(a) in Python's NumPy library can be used to produce power-law deviates, but *beware*, its argument a corresponds to $1 - \alpha$ under the power-law definition used throughout this book; see equation (B.7).

Another useful PDF is the Gaussian (or normal) distribution.[1] Gaussian distributions of random deviates can be easily generated

---

[1]Statistical theory would state that in the absence of cheating and with fair grading, and in the limit of infinite class size (ackpht!), the PDF of my midterm exam grades plotted in figure B.2 should be a Gaussian centered on the class average!

through the Box–Muller transformation, which produces *two* Gaussian deviates x_1, x_2 from *two* uniform deviates r_1, r_2 via the relations

$$x_1 = \sqrt{-2\ln r_1}\cos(2\pi r_2), \quad x_2 = \sqrt{-2\ln r_1}\sin(2\pi r_2),$$

$$r_1, r_2 \in [0, 1], \quad x \in [-\infty, \infty]. \tag{C.6}$$

The deviates so generated fill a Gaussian distribution of zero mean and unit variance. If the deviates need to be centered about a nonzero mean value (x_0, say) with a standard deviation $\sigma \neq 1$, then they should be rescaled as

$$g_1^* = x_0 + \sigma \times g_1, \qquad g_2^* = x_0 + \sigma \times g_2. \tag{C.7}$$

The function call g=numpy.random.normal(x0,sigma) in Python's NumPy library can be used to produce Gaussian deviates of mean value x_0 and standard deviation σ.

C.5 The Classical Random Walk

A random walk describes the changing position of an agent taking successive steps, all of the same length s, but oriented randomly, in the memoryless sense that, not only is the orientation of step n random, but it is also entirely independent of the orientations of previous steps.

Consider first a 1-D random walk, where the displacement is constrained to lie along a line (think of a very narrow road with high fences on both sides, or a very long doorless corridor within a building). The displacement at step n, measured with respect to some starting position, is denoted by D_n, and the two equiprobable steps are $s_n = \pm1$. By definition we can write

$$D_n = D_{n-1} + s_n, \quad n = 0, 1, 2, 3 \ldots. \tag{C.8}$$

Note already that the total distance walked, $n \times s$, is not the same as the displacement measured from the starting location; two steps to the right followed by two to the left add up to zero displacement, even though four steps have been taken. The squared displacement is then

$$D_n^2 = (D_{n-1} + s_n)^2 = D_{n-1}^2 + s^2 + 2D_{n-1}s_n. \quad \text{(C.9)}$$

Now consider a group of M agents, all starting at the same position and each engaging in a (collisionless) random walk. Introduce now the *ensemble average*, denoted by the brackets $\langle \ldots \rangle$, defined over this whole group:

$$\langle x \rangle = \frac{1}{M} \sum_{m=1}^{M} x(m). \quad \text{(C.10)}$$

Under this notation, the quantity $\langle D_n \rangle$ can be interpreted as the average displacement of the group as a whole. Averaging is a linear operator, in the sense that

$$\langle x + y \rangle = \langle x \rangle + \langle y \rangle, \qquad \langle ax \rangle = a \langle x \rangle, \quad \text{(C.11)}$$

where a is any numerical coefficient. Because of this linearity, applying our averaging operator to equation (C.9) yields

$$\begin{aligned}
\langle D_n^2 \rangle &= \langle D_{n-1}^2 + s^2 + 2D_{n-1}s_n \rangle \\
&= D_{n-1}^2 + \langle s^2 \rangle + 2\langle D_{n-1}s_n \rangle. \quad \text{(C.12)}
\end{aligned}$$

If no communication or interaction takes place between agents and consequently they have no way to get in step with one another, then for a large enough group of agents, $\langle s_n \rangle = 0$ since right- and left-directed steps are equiprobable. Moreover, for a memoryless process, the distribution of steps ± 1 at iteration n is entirely uncorrelated to the distribution of displacements D_{n-1} at the prior step. This implies

$$\langle D_{n-1}s_n \rangle = \langle D_{n-1} \rangle \langle s_n \rangle = 0. \quad \text{(C.13)}$$

This result is critical for all that follows. Equation (C.12) now becomes

$$\langle D_n^2 \rangle = \langle D_{n-1}^2 \rangle + s^2. \quad \text{(C.14)}$$

Setting $D_0 = 0$ without loss of generality, we have

$$\langle D_1^2 \rangle = s^2, \tag{C.15}$$

$$\langle D_2^2 \rangle = \langle D_1^2 \rangle + s^2 = 2s^2, \tag{C.16}$$

$$\langle D_3^2 \rangle = \langle D_2^2 \rangle + s^2 = 3s^2, \tag{C.17}$$

$$\vdots \tag{C.18}$$

and so, after n steps,

$$\langle D_n^2 \rangle = ns^2. \tag{C.19}$$

If the (discrete) variable n is interpreted as a temporal iteration, this expression indicates that the mean quadratic displacement increases linearly with time, so that

$$\sqrt{\langle D_n^2 \rangle} = s\sqrt{n}. \tag{C.20}$$

This is called the *root-mean-squared* displacement. It is important to understand that even though this increases with time, the mean displacement $\langle D_n \rangle$ vanishes at all times. The distinction is easier to understand by simulating a great many random walks and constructing distribution functions for the positions of the walkers. An example is shown in figure C.2, for a simulation involving 1000 1-D random walkers, all starting at $x = 0$. The distributions are constructed and plotted after 1000, 3000, 10,000, and 30,000 steps, as color coded. It is clear from these plots that the mean of each distribution, i.e., $\langle D_n \rangle$, always remains very close to zero, even after 30,000 steps, and all that time the most probable displacement, coinciding with the peak value of the PDF, is also essentially zero. Yet, equally obviously, the distribution spreads outward with time, so that the probability of finding a large displacement, either positive or negative, increases with time.

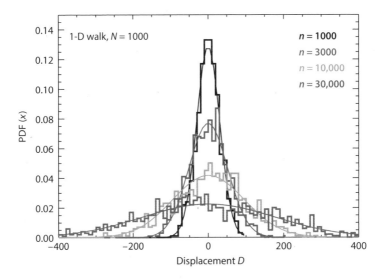

Figure C.2. Temporal spreading of the distribution of 1000 1-D random walkers, all originally located at $x = 0$. The thin lines are the Gaussian best fit to the distribution data, color coded correspondingly.

The colored thin lines are least-squares fits to these distributions, computed by adjusting the parameter σ of the Gaussian PDF

$$f(x) = \frac{1}{\sqrt{2\pi}\sigma} \exp\left(\frac{-x^2}{\sigma^2}\right). \qquad (C.21)$$

This parameter is a measure of the width of the Gaussian distribution (the full width at half-maximum is $1.176 \times \sigma$, with 68.3% of all measurements contained within $\pm\sigma$ of the mean). Here, σ can be shown to increase with time as $\sqrt{n}$, which is the same pseudo-temporal dependence arrived at when directly computing the root-mean-squared displacement (equation (C.20)).

All of these results carry over to random walks in more than one spatial dimension. The displacement $\boldsymbol{D}_n$ and step $\boldsymbol{s}_n$ become vector quantities, and equation (C.8) must be replaced by

$$\boldsymbol{D}_n = \boldsymbol{D}_{n-1} + \boldsymbol{s}_n, \quad n = 1, 2, 3 \ldots, \qquad \text{(C.22)}$$

where the step $\boldsymbol{s}_n$ still has unit length but is oriented randomly in space. The mean square displacement at step n becomes

$$\begin{aligned} D_n^2 &= (\boldsymbol{D}_{n-1} + \boldsymbol{s}_n) \cdot (\boldsymbol{D}_{n-1} + \boldsymbol{s}_n) \\ &= D_{n-1}^2 + s^2 + 2\,\boldsymbol{D}_{n-1} \cdot \boldsymbol{s}_n. \end{aligned} \qquad \text{(C.23)}$$

Once again, if the step orientation is truly random and uncorrelated to the displacement vector at the prior step, then averaged over a large ensemble of walkers, $\langle \boldsymbol{D}_{n-1} \cdot \boldsymbol{s}_n \rangle = 0$. Why this is so is exemplified in figure C.3, showing the first 18 steps of a 2-D random walk, beginning at $(x, y) = (0, 0)$, with a step length $|\boldsymbol{s}| = 1$, and the thick red line indicating the rms displacement vector after the 18th step, i.e., $\boldsymbol{D}_{18}$. This vector makes an angle θ_{18} with respect to the x-axis of a Cartesian coordinate system centered on $(0, 0)$, as indicated by the dotted lines. The 19th step will land the walker somewhere on the green circle of unit radius centered on the walker's position at the 18th step. Where on this circle the walker will actually land is entirely random, i.e., the angle α_{19} of its next step with respect to a local coordinate system centered on $\boldsymbol{D}_{18}$ can be anything between 0 and 2π, equiprobably. Now, the angle between the displacement vector $\boldsymbol{D}_{18}$ and $\boldsymbol{s}_{19}$ will be given by $\boldsymbol{s}_{19} - \boldsymbol{D}_{18}$, so that

$$\boldsymbol{D}_{18} \cdot \boldsymbol{s}_{19} = D_{18} s_{19} \cos(\alpha_{19} - \theta_{18}). \qquad \text{(C.24)}$$

At this point in the walk the angle θ_{18} is already set, at some value between 0 and 2π; whereas α_{19} is drawn randomly from a uniform distribution spanning $[0, 2\pi]$. Trigonometric functions being periodic, it is as if the angle $\alpha_{19} - \theta_{18}$ were also drawn from a uniform distribution spanning $[0, 2\pi]$; its cosine is therefore as likely to turn out positive as negative, both identically distributed, which ensures that an ensemble average of the above scalar product will always vanish—even though it almost never would

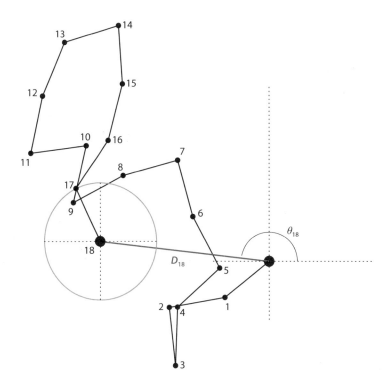

Figure C.3. The first 18 steps of a 2-D random walk in the plane, with unit-length step s_n. The numbered solid dots indicate the successive positions of the walker. The 19th step will land somewhere on the green circle centered on 18. Where it will land on that circle, i.e., the spatial orientation of that 19th step, is entirely independent of the length and orientation of the current displacement vector D_{18} (red line segment).

for a single walker. The same reasoning would also hold if both angles α_{19} and θ_{18} were drawn *independently* from their allowed range. This evidently also holds for any step n, and leads to the conclusion that the ensemble average $\langle D_{n-1} \cdot s_n \rangle = 0$. Therefore, the ensemble average of equation (C.23) becomes

$$\langle D_n^2 \rangle = \langle D_{n-1}^2 \rangle + s^2, \tag{C.25}$$

Random walk (100 steps)

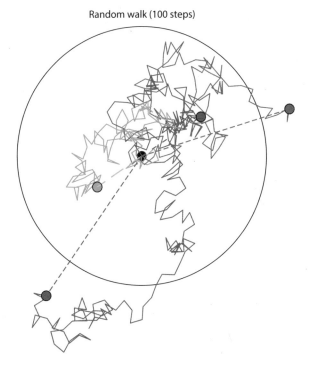

Figure C.4. Four 2-D random walks of 100 steps each. The starting point is indicated by a black dot at the center, and the circle indicates a displacement $D \equiv \sqrt{\mathbf{D} \cdot \mathbf{D}} = \sqrt{n} = 10$.

just as in the case of the 1-D random walk (cf. equation (C.14)). Everything else proceeds as before and leads again to equation (C.19). This is a truly remarkable property of random walks: no matter the dimensionality, the root-mean-squared displacement always increases as $\sqrt{n}$.

Figure C.4 illustrates a few 2-D random walks, each over 100 steps, with the circle drawn at the radius corresponding to the root-mean-squared displacement, $R = \sqrt{\langle D_{100} \rangle} = 10s$. This figure highlights, once again, the fact that the rms displacement is a statistical measure, obtained from an ensemble average over a very large number of walkers; the displacement of a given individual walker can deviate substantially from equation (C.25), while remaining bound in $[0, n \times s]$.

C.6 Random Walk and Diffusion

The gradual spreading observed in the distribution of random walkers in figure C.2 is prototypical of *diffusive processes*, and this turns out to be more than a mere visual analogy.

Let's stick to 1-D random walks and consider what happens at some arbitrary position x_0; only walkers within the range $x_0 - |s| < x < x_0 + |s|$ have a chance to cross x_0 at the next step, but then again, only if they happen to step in the required direction ($s = +1$ for walkers in $x_0 - |s| < x < x_0$, and $s = -1$ for those in $x_0 < x < x_0 + |s|$). Both stepping directions being equiprobable, on average only half the walkers on each side will then cross x_0. Denote by $\delta N(x_0)$ the net number of walkers crossing x_0 from the right to the left. This quantity will be given by

$$\delta N(x_0) = \tfrac{1}{2} N(x_0 - |s|) - \tfrac{1}{2} N(x_0 + |s|) \qquad (C.26)$$

(a negative value for δN would then mean that the net flow of walkers is from left to right). Now let $N(x_0, t)$ be the number of walkers standing somewhere in the full interval $x_0 - |s| < x < x_0 + |s|$ at time t. That number, at time $t + \Delta t$, will then be given by $N(x_0, t)$ plus the net number having entered from the left side at $x_0 - |s|$, minus the number having walked out to the right across the right boundary $x_0 + |s|$. Evaluating equation (C.26) at $x_0 - |s|$ and $x_0 + |s|$, instead of just x_0, then leads to

$$
\begin{aligned}
N(x, t + \Delta t) &= N(x, t) + \delta N(x - s) - \delta N(x + s) \\
&= N(x, t) + \left(\left(\tfrac{1}{2} N(x - 2s) - \tfrac{1}{2} N(x) \right) \right. \\
&\quad \left. - \left(\tfrac{1}{2} N(x) - \tfrac{1}{2} N(x + 2s) \right) \right) \\
&= N(x, t) + \tfrac{1}{2} \left(N(x + 2s) - 2N(x) + N(x - 2s) \right),
\end{aligned}
$$
$$(C.27)$$

where we have dropped the "0" index on x and the absolute value on s to lighten the notation. Dividing the right- and left-hand sides of this expression by Δt and rearranging terms, we get

$$\frac{N(x, t + \Delta t) - N(x, t)}{\Delta t} = \frac{1}{2} \left(\frac{(2s)^2}{\Delta t} \right)$$

$$\times \left(\frac{N(x + 2s) - 2N(x) + N(x - 2s)}{(2s)^2} \right). \quad (C.28)$$

Note that both the numerator and denominator of the right-hand side have been multiplied by the quantity $2s^2$. This mathematically legal, but by all appearances arbitrary, manoeuver was carried out so that the quantity within the second set of parentheses on the right-hand side is identical to a second-order centered finite difference formula for the second derivative of N with respect to x, with a spatial discretization increment $2s$. The term on the left-hand side is a first-order forward difference formula for the time derivative of N, with time step Δt. If these interpretations are accepted, then equation (C.28) can be viewed as a finite difference discretization of the partial differential equation

$$\frac{\partial N(x, t)}{\partial t} = D \frac{\partial^2 N(x, t)}{\partial x^2}, \quad (C.29)$$

where D is a *diffusion coefficient*, here given by

$$D = \frac{1}{2} \frac{(2s)^2}{\Delta t}. \quad (C.30)$$

Equation (C.29) is the well-known classical linear diffusion equation, which represents a macroscopic description of a random walk; it also describes the spreading of perfume (or other smell) in a room where the air is at rest, the slow diffusive mixing of cream in a coffee that is *not* being stirred, as well as a host of other common mixing and dilution processes. Equation (C.29) holds, provided the flux of the diffusing quantity is proportional to the (negative) concentration gradient of the diffusive substance, which is known as linear (or Fickian) diffusion. The physical link with the random walk arises from the random motion of perfume or cream molecules, continuously colliding with molecules making up the background fluid (air or coffee). Here is one of those instances where understanding the microscopic behavior, namely, the random walk, allows the macroscopic behavior, i.e., diffusion, to be calculated.

D

LATTICE COMPUTATION

Most computational implementations of the complex systems explored in this book are defined over lattices: sets of interconnected nodes on which the dynamical variables of the problem are represented. There are two interconnected concepts that need to be distinguished: lattice *geometry* and *connectivity*. Geometry is set by the relative positions of lattice nodes in physical space; connectivity refers to the coupling between nodes, i.e., which neighboring nodes interact with any given node. The foregoing discussion is framed in the context of 2-D lattices, but generalization to higher dimensions is usually straightforward.

When carrying our numerical simulations on lattices, nodal values are usually stored as arrays in the computer's memory, having the same dimension and lengths as the said lattices, i.e., nodal values on a 128×128 lattice are stored in a 2-D array having length 128 in each dimension. The syntax for defining such arrays is described in appendix A.2. From the user's point of view (but not in the computer's RAM), 2-D arrays are thus defined in terms of rows (first array dimension) and columns (second dimension), as with a matrix, which effectively represents a form of Cartesian geometry. Storing, accessing, and plotting nodal values for a Cartesian lattice is thus algorithmically trivial.

Lattice geometries other than Cartesian can still be stored in 2-D arrays, by a suitable choice of connectivity. The idea has already been illustrated in panels C and D of figure 2.5: with appropriate horizontal shifting of nodal positions, a 2-D Cartesian lattice with anisotropic 6-neighbor connectivity can be reinterpreted as a triangular lattice with isotropic 6-neighbor connectivity. Geometry is secondary (except when plotting!), and connectivity is the key.

D.1 Nearest-Neighbor Templates

Setting connectivity for a lattice is best accomplished using a *nearest-neighbor template*. This gives the *relative* positions of nearest neighbors with respect to a given nodal position (i, j). For example, on a Cartesian lattice the 4 nearest neighbors of a node (i, j) are located at the bottom, right, top, and left; or, in terms of nodal numbering, $(i + 1, j)$, $(i, j + 1)$, $(i - 1, j)$, and $(i, j - 1)$. This can be stored in two 1-D integer arrays of length 4, one for each lattice dimension. Under Python/NumPy this is initialized as follows:

```
dx=np.array([1,0,-1,0])
dy=np.array([0,1,0,-1])
```

This is known as the *von Neumann neighborhood*. Including the 4 diagonal neighbors yields the *Moore neighborhood*, which would be defined by the template arrays

```
dx=np.array([0,1,0,-1,-1,1,1,-1])
dy=np.array([-1,0,1,0,-1,-1,1,1])
```

For the 6-neighbor triangular lattice of figure 2.5, the template arrays are

```
dx=np.array([0,1,0,-1,-1,1])
dy=np.array([-1,0,1,0,1,-1])
```

Whatever the connectivity, lattice operations can use these template arrays to efficiently retrieve nearest-neighbor information. For example, with nodal values stored in a 2-D array named grid, calculating the sum of nodal values for all nearest neighbors of node (i, j) under the von Neumann neighborhood could be coded like so:

```
sum_nn=0.
for k in range(0,4): sum_nn+=grid[i+dx[k],j+dy[k]]
```

or equivalently, by invoking the sum() function from Python's NumPy library:

```
sum_nn=np.sum(grid[i+dx[:],j+dy[:]])
```

These instructions would be typically embedded within two loops for the indices i and j, thus scanning all lattice nodes. There is one pitfall to this strategy: as shown in figure D.1, still for the von Neumann neighborhood, it will fail for nodes at the boundaries of the lattice, which have only 3 nearest neighbors (and only 2 for the four corner nodes).

There are ways out of this difficulty, of course. In principle, the most straightforward is to treat boundary nodes separately, for example through the use of suitably modified template arrays used only for boundary nodes. However, this leads to cumbersome extra coding that significantly lengthens a simulation code and reduces its readability. A better strategy is to make use of *ghost nodes*, as shown in figure D.2. The 10 × 10 lattice of figure D.1 is now padded on all sides with a layer of additional nodes (open gray circles). This expanded lattice is now of size 12 × 12, but computations associated with the model's dynamical rules take place only in the interior 10 × 10 block of nodes corresponding to the original, unpadded lattice. Unlike in figure D.1, using the 4-neighbor template on the red node, now numbered

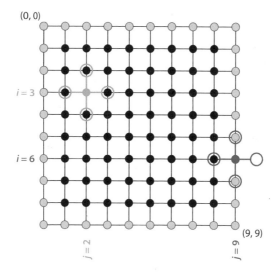

Figure D.1. Use (green) and misuse (red) of a 4-nearest-neighbor template on a 10×10 Cartesian lattice. Nodes are numbered by a pair of indices (i, j), starting at the top-left corner and increasing downward and to the right. The template functions well for interior nodes (black) but will lead to out-of-bounds array indexing for boundary nodes (gray) unless alternative, reduced template arrays are introduced for boundary nodes, or additional conditional instructions (**if. . . else**) are added within the code.

$(i, j) = (7, 10)$, will not exceed the array length in the horizontal since $j + 1 = 11$, which is now legal.[1]

Which numerical value is to be assigned to ghost nodes is dependent on the boundary conditions of the problem. In the earthquake model of chapter 8, for example, the ghost nodes are

---

[1] Some computing languages allow the use of negative integers to index array elements, so, for example, here each dimension of the 12×12 lattice could have the nodes numbered from -1 to 10, so that the green and red nodes retain their original numbering $(i, j) = (3, 2)$ and $(6, 9)$, as in figure D.1. I stayed away from this cleaner numbering strategy for reasons of portability to languages that do not allow such generalized array indexing.

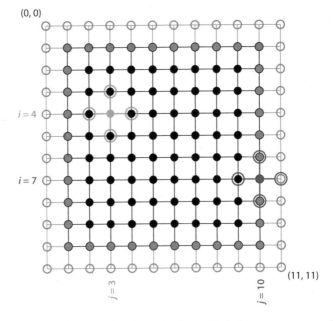

Figure D.2. The same lattice as in figure D.1, but now surrounded by a layer of ghost nodes (open gray circles). The full lattice is now of dimensions 12 × 12, and the 4-neighbor template can be used on all nodes of the embedded original 10 × 10 lattice even for the (true) boundary nodes (solid gray circles).

simply set to zero and retain that value throughout the whole simulation. This is as easy as it gets. In the hodgepodge machine simulations of chapter 11, on the other hand, the ghost nodes are used to enforce *periodic boundary conditions*, as detailed in the following section.

D.2 Periodic Boundary Conditions

In some lattice-based models introduced in this book, periodic boundary conditions are imposed on the lattice. Sometimes this is dictated by the geometry of the problem. Consider, for example, ants walking on the surface of a sphere. Using latitude–longitude coordinates, an ant walking eastward and crossing longitude $360°$ must instantly "reappear" at longitude $0°$ because both correspond to the same point on the sphere. Longitudinal periodicity is then mandatory. Now imagine ants walking on the surface of a torus. Longitude is again periodic, but now, so is the latitudinal direction, since an ant starting in the equatorial plane and walking north will travel a circular path that will bring it back to its starting point. When simulating such a walk, the torus can thus be mapped to a square with periodicity enforced both horizontally and vertically. This is the geometric interpretation to be ascribed to the highway-building ant of section 2.4, and to the flocking simulation of chapter 10.

In other instances, periodic boundary conditions are used simply because we cannot specify boundary values, and doing so arbitrarily would perturb the evolution of the system. This is the case with the hodgepodge machine simulations of chapter 11. Enforcing periodic boundary conditions then implies that the simulated domain is but a "tile" that repeats itself across space to infinity, exactly like on a tiled floor (made of identical tiles, and without the infinity part). In such a situation, we must simply accept the fact that the simulation cannot generate or accommodate structures that have length scales larger than the simulated periodic unit.

Figure D.3 shows how ghost nodes can be used to enforce periodic boundary conditions. The true boundary nodes of our now familiar original 10×10 lattice have been colored according to the side they belong to, with four more distinct colors used for the corner nodes. If the original lattice were to be replicated horizontally and vertically to tile the whole space under the assumption of periodicity (as for the unit square domain in figure 10.4), then the top row of ghost nodes (open blue circles plus corner nodes) are really the "same" nodes as the bottom row of the original 10×10 lattice (solid blue nodes plus corner nodes, boxed in blue). Vertical periodicity can therefore be enforced by

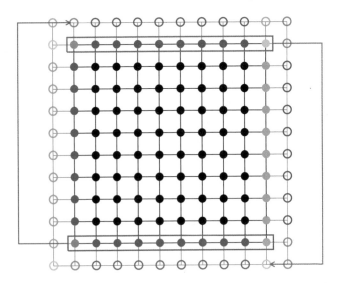

Figure D.3. Enforcing periodic boundary conditions via ghost nodes. The boundary nodes of the original 10×10 lattice (solid gray nodes in figure D.1) are copied to the ghost node layer (open circles) on the opposite side of the enlarged 12×12 lattice, following the color coding given. Note how the corner nodes of the 10×10 lattice get copied into three distinct ghost nodes. Interior nodes (in solid black) remain unaffected by this whole procedure.

copying the 10 nodal values of this row (boxed in blue) to the corresponding nodes of the top row of ghost nodes, as indicated by the blue arrow. The same applies to the top row of the original 10×10 lattice (purple), which gets copied into the bottom row of ghost nodes. The same procedure is used in the horizontal direction, as indicated by the color coding on the figure. Note how each corner node in the 10×10 lattice gets copied thrice into ghost nodes: once horizontally, once vertically, and once diagonally to the opposite corner of the 12×12 lattice. The function `periodic`, given within the code in figure 11.4, gives a compact algorithmic implementation of this procedure.

Periodicity in one spatial dimension amounts to assuming that the 1-D domain is a closed ring, which is much easier to implement; see, for example, the 1-D CA code of figure 2.4.

D.3 Random Walks on Lattices

Random walks can be defined over a lattice, with walkers constrained to move from one node to a randomly selected nearest-neighbor node, according to some suitably defined neighbor template. All that is needed is to generate a random integer to pick an element of the appropriate template arrays. The following code fragment shows how to set up a random walk of N steps, here on a 2-D Cartesian lattice with 4-neighbor connectivity, and with the walker starting at the (arbitrary) nodal position $(i, j) = (5, 5)$ on the lattice:

```
1   import numpy as np
2   N=100                        # number of random walk steps
3   dx=np.array([0,1,0,-1])      # template arrays for 4 neighbors
4   dy=np.array([-1,0,1,0])
5   ...
6   i,j=5,5                      # initial nodal position of walker
7   for  k in range(0,N):       # walk N steps
8       r=np.random.choice([0,1,2,3])  # random integer between 0 and 3
9       i+=dx[r]                 # take one random step
10      j+=dy[r]
11  ...
```

An example is shown in figure D.4. The first 18 steps of the walk are shown, and reveal some occasional backtracking (steps 2–3–4 and 13–14–15). Note also that, after the sixth step, the walker is actually back at its starting position.

This may appear to be a strongly constrained type of random walk, but when simulating many such random walks on the lattice over a great many steps, the orientation of the displacement vector is effectively random, and its ensemble averages $\langle \boldsymbol{D} \rangle_n = 0$, and $\langle D^2 \rangle \propto n$, just like in a classical random walk (see section C.5).

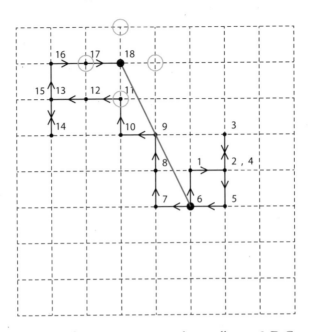

Figure D.4. The first 18 steps in a random walk on a 2-D Cartesian lattice with 4-neighbor connectivity, starting at the lattice center. Successive nodal positions are numbered, and the thick red line segment indicates the displacement vector $\boldsymbol{D}_n$ at step 18. The 19th step will land the walker on one of the four nodes circled in green, which one being chosen randomly, in a manner independent of the current position or direction of past steps (see text). Compare to figure C.3.

Moreover, on length scales much larger than the internodal distance, the spatial distributions of walkers are essentially the same in both cases. Figure D.5 illustrates this, now for four 400-step random walks on a larger lattice, still with 4-neighbor connectivity. For displacements much larger than the microscopic scale set by the internodal distance, the spatial distribution of end points is statistically undistinguishable from that associated with a conventional 2-D random walk (cf. figure C.4).

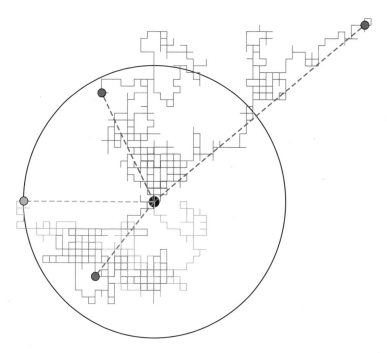

Figure D.5. Four 400-step random walks on a lattice with 4-neighbor connectivity. The starting point is indicated by a black dot at the center, and the circle indicates a displacement $D \equiv \sqrt{\boldsymbol{D} \cdot \boldsymbol{D}} = \sqrt{n} = \sqrt{400} = 20$. Compare to figure C.4.

Random walks on lattices become particularly useful when simulating a system where many walkers moving simultaneously on the lattice interact locally in some way (e.g., the healthy and sick agents in the epidemic propagation simulations of chapter 9). Knowing the position (i, j) of a walker on the lattice, only this node (or nearest-neighbor nodes) must be checked for the presence of another walker. In a classical random walk, this would involve instead the calculation of $\simeq N^2/2$ pairwise distances, to pick which walkers are within some set distance inside which the interaction takes place (as for the repulsion and

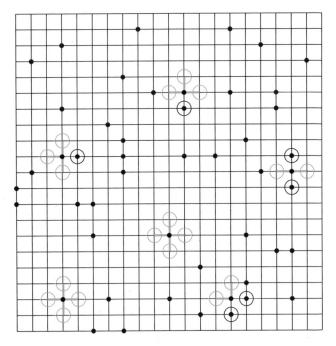

Figure D.6. Displacement rules for random walks on a lattice in which walkers are not allowed to move to a node already occupied by another walker. Here, 50 walkers (black solid dots) are distributed on a 20 × 20 lattice. For a subset of 6 walkers, the four target nodes are indicated by green circles, and the allowed moves by thick green line segments.

flocking forces in the flocking simulations of chapter 10). This becomes computationally prohibitive for very large N. There exist strategies and algorithms to reduce this number, but they are too complex (!) to get into even this book.

Another, related attractive feature of random walks on lattices is the possibility of accommodating simplified representations of "collisions" between two walkers. The idea is illustrated in figure D.6. Often, at a given temporal iteration, only a small fraction of lattice nodes are occupied by random walkers (black solid dots). Every one of these would normally take its next step to one of the four possible positions indicated by the stencil of four green circles centered on each walker, as indicated in the figure for just six walkers. The idea is to void the step if it would land the walker on a lattice node already occupied by another walker; in such a situation, the walker remains on their node until a new random step is attempted at the next temporal iteration. The allowed steps for the six selected walkers in figure D.6 are indicated by the thick green line segments. The fact that walkers cannot move to an occupied node represents a form of collision, since two walkers on neighboring nodes cannot cross but instead tend to move away from each other, in a statistical sense. The flow of fluids can be simulated in this manner. Chapter 8 of Wolfram's book on cellular automata (cited at the end of chapter 2) presents a few nice examples. See also the Wikipedia pages on *lattice gas automaton* and *lattice Boltzmann methods* for entry points into the technical literature:

http://en.wikipedia.org/wiki/Lattice_gas_automaton;
http://en.wikipedia.org/wiki/Lattice_Boltzmann_methods.

INDEX

Agent Smith, 242, 253
agents, 41, 51
 active vs. passive, 240, 243, 246
 ant, 41, 51, 103
 as car drivers, 154
 as flockers, 225
 termite, 50
angle of repose, 106, 115, 125
artificial life, 51
avalanches, 108
 in epidemic model, 199
 falloff, 117
 in forest-fire model, 131, 137, 143
 in OFC earthquake model, 180, 184
 in sandpile model, 108, 118, 121
 size measures, 119, 121, 137, 143, 184, 195
 in traffic model, 164

Bak, Per, 129
bifurcation diagram, 12
Black Death, 198, 204, 213, 222
boundary conditions
 closed, 110, 113, 133, 201
 open, 110, 113, 192
 periodic, 31, 43, 151, 228, 232, 260, 342–344

bozos, 155, 168, 170, 242
brute force, 4, 235
Burridge–Knopoff stick–slip model, 175–178, 196

CA, *see* cellular automata
cardiac arythmia, 266
cellular automata, 23
 1-D vs. 2-D, 31
 classes, 30
 connectivity, 32
 hodgepodge machine, 254
 probabilistic, 130
 rules, 29, 38
 simulating fluid flows, 349
 two state, 23
chaos, 10, 21
chemical oscillations, 249–251, 274
clusters
 in forest-fire model, 135, 137
 in percolation, 82, 91, 99, 130
 in traffic model, 167
 self-similarity, 91
 tagging algorithm, 86–90, 105, 134, 166
complexity
 and consciousness, 290
 and the origin of life, 291